Kirill Vasilchenko

RECURSIVE SPIRIT

A Psychiatrist Explains Why Formulas and Spirits Are the Same Thing, and Why This Knowledge Makes You Free and Lonely

Recursive Spirit

A Psychiatrist Explains Why Formulas and Spirits Are the Same Thing,
and Why This Knowledge Makes You Free and Lonely

Independently Published

First edition, 2026

Recursive Spirit

Kirill Vasilchenko

To my family —

who found the strength, from somewhere,

to endure my mental escapism

without expressing a drop of displeasure.

We walked this path together,

though the path was mine.

This, quite possibly, is what reduced

the probability of my hospitalization.

Author's Note

This book was originally written in Russian. The English edition is not a translation but an adaptation: the argument, the stories, and the science are the same, but the voice, the syntax, and certain culturally specific passages have been rebuilt for the English-language reader. The Russian original is available under the same title.

Preface

This book is about processes that cannot be seen, cannot be touched, cannot be pinned to a location — yet are real in the most practical sense of the word: they make things happen. They shape how people behave, determine which configurations of life are stable and which fall apart, and they include the observer as part of themselves. These processes don't fit inside your head — not because you're not smart enough, but because the instrument you're using to think about them is itself part of what you're trying to think about. There is a ceiling, and it is built into the architecture.

It follows that working with these processes requires intermediaries — interfaces. There are two types. One is formulas, models, equations: the language of science. The other is spirits, entities, agents: the language used by the ancients, by shamans, by mystics, and by Castaneda. Both languages describe the same thing. But the second one engages computational resources unavailable to the first — which is why a description in the language of spirits sometimes predicts more accurately than a description in the language of formulas.

Most people live inside one of these languages and mistake it for reality. The scientist is certain that the world is formulas. The believer is certain that the world is spirits. Both are locked inside their map, not knowing it's a map.

Freedom — and this is what this book is for — arrives when you see that both languages are maps. That a formula is not the world but a description of the world. That a spirit is not an entity but an interface. And that you can switch between them without confusing either one with reality. Freedom is not knowledge. Freedom is *the ability to choose your language*.

This freedom comes with loneliness, because the person who sees both maps cannot fully belong to either world — not among the scientists, not among the mystics. And it cannot be explained, because explaining it requires the other person to see both maps too. They see one. They don't want to see the second. Not because they're stupid — because they don't need to.

* * *

I am a psychiatrist — someone whose profession consists of distinguishing a working mind from a broken one. But a psychiatrist with a caveat that needs stating upfront, or the book will be misread.

I am a psychiatrist by training and by the years I've given to it, years I continue to give. I respect my colleagues — every one of them — because this profession demands a kind of courage no test can measure. But I do not accept the version of psychiatry that currently exists. Not because I don't love it — because I can see it standing still. I have sat through countless case conferences where a team of experienced clinicians would spend hours dissecting a patient's phenomenology, arrive at a plan with visible relief — and then, weeks or months later, the attending physician would quietly change the treatment because it wasn't working. Case conferences, I should note, are a particularly elaborate tradition in the psychiatric training I went through — marathon sessions of collective phenomenological analysis that can be intellectually thrilling and clinically useless in roughly equal measure. But the underlying problem is universal: psychiatry, everywhere, still diagnoses by clinical impression and treats by trial and error, cycling through medications like keys on a ring until one of them fits. It is 2026, and we are still doing this.

I am a psychiatrist with a heavy IT lean — data science, neural networks, computational methods. Everything I do now is an attempt to build a bridge between computational technology and psychiatry. Not the psychiatry that exists today, but the one that will come. For that, at least one generation needs to turn over — psychiatrists raised on algorithms and data the way the current generation was raised on phenomenology and clinical intuition. Spending hours dissecting a patient's inner world, mapping their experience in phenomenological detail — I enjoy this, I do it, I'm trained in it. But I'll say it plainly, and I expect the same honesty from my colleagues: without computational tools, it is not much more than intellectual masturbation. Beautiful, stimulating, giving the feeling of depth — but not advancing the ball one inch. That's a separate book, though. In research, this is exactly what I work on — my publications address precisely this problem. The rest of my biography unfolds in the chapters of this book.

* * *

So: I read Castaneda, didn't believe a word of it, started finding confirmations in neuroscience and clinical practice — and was forced, at every new discovery, to run a differential diagnosis on myself: is this an insight, or am I becoming my own patient?

The book is the story of that diagnosis. It begins with breathing and the word 'spirit,' passes through neural networks, philosophy of mind, rehab centers, rocket attacks, and night shifts as a security guard with a phone and Castaneda — and ends with an answer to the question the whole journey was for: what distinguishes a mage from a paranoid.

The answer: calibration. The ability to stop, check, and keep going. Or not keep going — if the check says don't.

* * *

A warning — for those it might bother: this book contains profanity. It appears where it's needed, for emotional precision and because there is no other way to say it. As one of the chapters argues, some languages describe certain things better than all others. In the Russian original, the profanity is more pervasive — Russian *mat* is not quite the same as English swearing; it's a separate linguistic register with its own grammar and emotional architecture. I've toned it down for this edition, but not eliminated it. 'Psychotic decompensation' is a precise term, but 'you cracked' conveys something the term does not. If this puts you off — close the book, take your sensibilities elsewhere. I won't be offended.

I have a weakness for combining things that aren't supposed to go together. Academic language and profanity. Neural networks and Castaneda. Formulas and spirits. Imagine how much livelier conferences would be if a professor — not to insult anyone, just for emphasis — said: 'Colleagues, this is, to put it gently, a clusterfuck. This cannot stand.' It holds attention. It delivers the message faster. Who knows, maybe those interminable case conferences would move along quicker. But there are games you have to play in academia.

Speaking of games. Originally, I planned to write an academic literature review explaining the problem of two interfaces. And honestly, I'd already started. But one of the things I despise about academic life is the endless rounds of peer review, where the reviewer typically attacks the formatting rather than the substance. Alongside this book, I'm working on a paper that is essentially a mathematical field of experiments confirming the theses laid out here. It will come. But the genre of academic writing is a sedative — it puts me to sleep even when I'm writing, let alone reading. So I decided to do something more fun. For the first time in my life, I'm attempting a work that is literary but grounded in autobiography and real scientific data — a book anyone can read. It's simply faster and more accessible this way.

* * *

If you're holding this book, you are probably one of two kinds of people. Either you've encountered something that works even though it shouldn't — and you want to understand why. Or you're simply curious what happens when a psychiatrist reads Castaneda and doesn't lose his mind.

Either way — welcome.

I wrote this book for myself fifteen years ago. For the person who noticed something strange and didn't know what to do with it: believe it or treat it. If you are that person, this book is for you. If you're not — read it anyway. Worst case, you'll say 'that's really interesting' in the tone of voice I'll tell you about in the final chapter.

Contents

Part I
Rustling in the Bushes

How I noticed that explanations were working even though I didn't believe them. And how, at every discovery, I was forced to check: is this an insight — or am I becoming my own patient?

Chapter 1. Breath

How I noticed that explanations were working even though I didn't believe them.

And how, at every discovery, I was forced to check:

is this an insight — or am I becoming my own patient?

Opening move. The history of the word 'spirit' as the history of the first interface.

A person dies, and the room changes.

Everything about the change is measurable. Body temperature drops at a predictable rate — roughly a degree per hour, adjusted for mass and ambient temperature. Pupils dilate and fix. The jaw goes slack. Blood pools in the lower tissues, leaving the livid patches that forensic pathologists call livor mortis. Rigor sets in after a few hours. All of it can be quantified, graphed, published.

But if you have ever been in the room when it happens — not as a medical examiner cataloging data, but as a person standing next to someone who was talking to you thirty seconds ago — you know that the measurable changes do not describe the event. Something leaves. Not metaphorically. Not poetically. Something that was there ceases to be, and the body that remains is not the person who was breathing half a minute ago. It is an object in the shape of a person. The difference is absolute, and no instrument in the hospital registers it.

I observed this in clinical settings more times than I care to count. I am a psychiatrist — my specialty is the mind, not death — but before residency I worked as an orderly in intensive care, and death has a way of finding you even when you're not ready. When the first patient died in my presence, I thought nothing deep. I thought: remove the oxygen mask. Update the chart. Training teaches you to think in protocols when the alternative is to feel something you're not yet equipped to feel.

It came later. On the walk home, in the November darkness. Not as a thought — as a physical sensation: a sudden awareness of air entering and leaving the lungs, cold stinging the nostrils, small clouds of vapor at each exhale.

Breathing. The person had stopped breathing, and the person had stopped being.

This observation is roughly a hundred thousand years old. And it is where everything in this book begins.

* * *

I did not begin this book with Castaneda. Nor with cognitive science, philosophy of mind, or the elaborate theoretical architecture that will occupy the following chapters. I began the way every human who has ever thought about this began: watching someone die and noticing that what left could not be named.

This is the oldest observation in the world. Older than writing. Older than agriculture. Probably older than language — because you don't need words to notice: breathing means alive, not breathing means dead. A wolf notices this. A crow notices this. The observation is not human — it is mammalian, possibly vertebrate. What humans added was language. We gave it a name.

And here is the first remarkable thing: in languages that never touched each other, in civilizations separated by oceans and millennia, we named it the same way.

But before I show you how — a small digression. Because my ability to notice what I'm about to show you has a specific origin, and it's not in the library of cognitive sciences.

In my first year of medical school, one of my favorite subjects was Latin. Not Latin itself — I tolerated the grammar and despised the declension tables. But the department head had an obsession: Latin had to be taught through etymology. Not rote paradigms — the history of words. Where did the word come from? What did it mean before it became a medical term? How did its meaning shift? This was not standard methodology, and the other faculty thought he was eccentric. He was eccentric. But he planted something in me — a reflex to look at a word and ask: *what did you used to be?*

And now — something slightly embarrassing. About two months into first year, a rumor went around the faculty: a lecturer had hurled a set of keys at a student's head in the middle of class. I heard this third-hand and belatedly realized the student was me. I'd been called to the board to decline a word. I couldn't. The lecturer — a small, fierce woman who, I later learned, had once been an excellent classical philologist before taking up teaching medical Latin — lost patience and threw her keys at me. They missed. But the word she asked me to decline was the word I would spend a decade trying to understand.

Spiritus. Fourth declension, if anyone cares. I've since learned it. But the real declension — the one this book works through — I am still declining.

* * *

Ruach (רוּחַ). Hebrew. The word appears roughly four hundred times in the Old Testament. It means 'wind.' It means 'breath.' It means 'spirit.' Not three different words that happen to sound alike — one word with one meaning, for the conveyance of which English needs three. In the second verse of Genesis, *ruach elohim* moves over the face of the waters — and the translators have argued for centuries whether this is the 'Spirit of God,' the 'breath of God,' or the 'wind of God.' The argument exists because in Hebrew there is no argument. It is all three. Simultaneously. Without contradiction.

Pneuma (πνεῦμα). Greek. Air in motion. Wind bending trees. Breath filling lungs. Spirit animating a living being. In the Gospel of John, Jesus tells Nicodemus: 'The *pneuma* blows where it wishes, and you hear its sound, but you do not know where it comes from or where it goes.' The sentence works because in Greek it is simultaneously about wind and about spirit. No metaphor needed. A single word covers the territory that English breaks into pieces.

Ātman. Sanskrit. The word that became the philosophical foundation of Indian metaphysics — the eternal, unchanging Self at the center of Hindu philosophy — began its career as the word for breathing. The etymology is impeccable: from Proto-Indo-European *h_1éh_1t-mō — 'breath,' a cognate of Old English *æþm* ('breath'), German *Atem* ('breathing'). The most abstract concept in Indian philosophy grew from the most concrete bodily observation.

Qì (氣). Chinese. The character consists of the radical for 'vapor' and the sign for 'rice' — steam rising from cooking grain. In the *Lunyu* it simply meant 'breath.' By the Han dynasty it had become the fundamental substance of all reality — simultaneously material and energetic, physical and vital. You cannot separate the physical from the spiritual in *qì*, because Chinese thought never performed that separation. The word wouldn't allow it.

And the Russian word, which the reader of the original edition uses every day without thinking: *dukh*. From Proto-Slavic *duxъ, from Proto-Indo-European *dhew- — 'to breathe, to blow.' Spirit. Breath. Soul. Air. Gust. All from one root.

Latin *spiritus* — from *spīrāre*, 'to breathe.' Direct ancestor of English *spirit*, as well as *inspire* (to breathe into), *expire* (to breathe out — and to die; note the coincidence), *respiration*, and *aspiration*. Every time you use the word 'spiritual,' you are saying 'breath-ual.' You just don't notice.

* * *

I want to be precise about what this pattern proves

and what it does not. It does not prove that spirits are real. It does not prove that the ancients had access to mystical truth. It supports no New Age thesis about the wisdom of premodern cultures or the spiritual bankruptcy of modernity.

It proves something more interesting. And more unsettling.

These languages never contacted each other. Proto-Indo-European *h_1éh_1t-mō and Chinese qì share no historical point. Hebrew ruach and Sanskrit ātman evolved on different branches of the human tree. The convergence is not cultural diffusion. Not transmission of knowledge from one civilization to another. It is independent invention — on at least four continents, in at least five unconnected linguistic traditions — of the same conceptual identification: breath = wind = spirit.

The question is: why?

And the answer — the one that began to change everything I thought about the human mind — is that these languages were not imprecise. They were more precise than us.

* * *

Here I need to put on a different hat for a moment — the academic one, the one that smells of peer-reviewed journals and pinches at the temples. Bear with me. What comes after is worth it.

In cognitive science there is a concept called the Hyperactive Agency Detection Device, or HADD — a term coined by Justin Barrett, building on earlier work by

Stewart Guthrie. The idea is simple: the human brain is wired to detect agents. Moreover, it is overwired. We see faces in clouds, intentions in coincidences, purpose in random noise. The evolutionary logic is straightforward — if you see a shadow in the bushes and assume it's a predator, you lose nothing if you're wrong and survive if you're right. If you assume it's just a shadow and you're wrong, you're lunch. Natural selection favors the paranoid. Those who heard rustling and thought 'someone is there' left more descendants than those who thought 'it's just the wind.'

This is textbook. Any introductory course in cognitive science of religion covers it, usually with a slide of a blurry shape in the bushes and the caption: 'Leopard or rock?' The answer, as far as your amygdala is concerned, is always: 'Leopard.' Better safe than digested.

But there is a subtlety that most textbooks miss, and it changes everything.

Not all ambiguous stimuli are created equal. A shadow on the ground is a weak trigger for agent detection — you need a specific shape, context, lighting. But breathing? Wind? In cognitive terms, these are ideal triggers. Here's why.

Breathing has four properties that no other natural phenomenon combines. First, universality: every human observes breathing from birth, no special knowledge required. Second, correlation with the most significant event in any human's experience: death. Stopped breathing — dead. This is not a statistical correlation that requires a sample and a p-value. It is absolute, observed in every case, without exception. Third, invisibility: the breath itself cannot be seen. You see its *effects* — the chest rising, the vapor in cold air, the sound — but the thing itself is transparent. Fourth — and this is where it gets interesting — agency. Breathing looks like something that *does* things. It enters and exits. It can be forceful or gentle. It responds to emotional states. It behaves, in the most literal sense, as if it has volition.

George Lakoff and Mark Johnson, in their work on embodied cognition, would call this a primary metaphor — a pre-conceptual identification between sensorimotor experience and an abstract domain, formed automatically through lived experience, long before any philosophical reflection. The correlation between breathing and life does not need to be taught. It inscribes itself.

And when, a hundred thousand years ago — or two hundred, or fifty, the dating doesn't matter — a human stood over the body of someone who had just died and noticed that breathing had ceased, the cognitive machinery did exactly what it was designed to do. It detected an agent. Invisible, powerful, unlocalized, and causally effective — because a moment ago this body was alive, and now it is not, and the only visible change is that the breathing stopped. Something left.

That agent was the first spirit.

And the word for it was the word for breath.

* * *

The standard modern reading of this story goes something like this: the ancients got confused. They saw a correlation between breathing and life, and since they lacked scientific understanding, they made a mistake — they attributed agency to a natural process. They invented spirits. Spirits do not exist. End of story. Progress marches on.

Edward Tylor, the Victorian anthropologist who first mapped this territory in 1871, was more generous but still condescending. He called the people who made this conceptual move 'ancient savage philosophers' and, while admiring their reasoning, framed it as a primitive error — the first step in a long march from superstition to science, from dark to light, from confusion to clarity.

I believed this story for a long time. It is a good story. Clean, progressive, flattering to everyone who believes it.

It is also wrong. More precisely, it is exactly half right, and the half where it's wrong is the interesting half.

The 'ancient savage philosophers' were not confused in their observation. They saw something real: an invisible, unlocalized process with visible, causally effective manifestations. Something that exists wherever conditions permit and nowhere in particular. Something you cannot point to, cannot hold, cannot weigh — but whose absence means death. They named this observation with the most precise word available to them. One word. For breathing, wind, and spirit. Because to them — and here is the key — *these were not three different things.*

Their description of this observation — one word for breath, wind, and spirit — is not imprecise. It is, in fact, more precise than anything we have produced in the two thousand years since. Because they had not yet made Plato's mistake — the one we inherited. They had not yet split the world into matter and spirit, body and soul. They worked with an intact vocabulary that tracked something real: an invisible, substrate-independent, causally effective process continuous with the observer.

This description — strip away the ancient lexicon and substitute modern terms — is a remarkably good description of what computational science calls an algorithm. Or what Gregory Bateson called 'the pattern which connects.' Or what this book will call an immanent pattern.

But I'm getting ahead of myself.

* * *

The closest anyone came to what I am trying to articulate in this book was not the mystics. It was a school of philosophers who drank cheap wine and lectured under a painted colonnade in the Athenian agora.

The Stoics took the word *pneuma* — breath, wind, spirit — and turned it into what may be the most sophisticated theoretical physics of the pre-scientific era. For the Stoics, *pneuma* was not a metaphor. It was an active, organizing principle of the entire cosmos: a mixture of fire and air pervading all matter and determining its properties through internal tension — *tonos*. Different degrees of pneumatic tension produced different levels of organization: *hexis* (cohesion) in stones, *physis* (nature, growth) in plants, *psyche* (soul, animation) in animals, and *logikos* (rational soul) in humans. Not four different substances — one process at four levels of complexity.

The physicist Samuel Sambursky showed in 1959 that Stoic *pneuma* functioned as a field theory two millennia before field theory existed: a continuous, tensile medium pervading all things, whose internal dynamics determined the properties of every region it occupied.

If you are a philosopher of mind, the term for this is 'substrate-independent.' If you are a programmer, you would say 'algorithmic.' And if you were a Stoic standing under the painted colonnade with a cup of rough wine, you would say: *pneuma*.

The word itself — breath — preserves the original intuition. Breath does not belong to any particular body. It is not made of lung tissue or blood. It passes through bodies, animates them, and moves on. The same air in your lungs an hour ago was wind in the trees. The ancient vocabulary was not confused. It was tracking a real feature of reality with the most precise instrument available: a single word that refused to make distinctions the phenomenon itself does not make.

* * *

And then Plato broke it.

I'm being slightly unfair. What Plato did was brilliant and in some sense necessary. But for the story this book tells, his intervention was a catastrophe — not because it was wrong, but because it was so convincingly right in certain respects that it became impossible to see where it was devastatingly wrong in others.

Before Plato, *psyche* was breath. After Plato, *psyche* was an immortal rational soul — an entity fundamentally different from the body, capable of existing without it, possessing abstract knowledge through recollection (*anamnesis*). The physical tether was severed. The word kept its old skin but inside was a completely different animal.

In the context of this book, Plato's move was a loss of precision. The pre-Platonic identification of spirit with breath preserved a key intuition: the animating principle is not a separate substance but a process inseparable from the system it animates. You cannot extract breathing from a body and examine it on a table. It exists only as long as it is happening, only inside the system where it occurs. The moment you remove it — the system dies. This is a remarkably accurate description of what modern computational theory calls a substrate-dependent process — or, more precisely, of a process that is substrate-independent in principle but immanent in practice: always realized in some substrate, never identical to any particular one.

Plato's innovation was to extract the soul from the body and declare it independent. Immortal. Separable. A thing, not a process. This was the birth of Western dualism — the split between consciousness and matter, soul and body, spirit and world — and it remains the operating system of Western thought twenty-four centuries later, even among people who consider themselves materialists. When a neuroscientist says 'consciousness is produced by the brain,' that sentence assumes the Platonic architecture — subject and object, producer and product, that which observes and that which is observed. Plato is built into the grammar.

Translations completed the job. When the Hebrew scriptures were translated into Greek in the third and second centuries BCE, the translators of the Septuagint rendered *ruach* as *pneuma* and *nephesh* as *psyche* — but by then the Greek words had already been Platonized. They carried dualistic baggage absent from the Hebrew originals. The Hebrew *ruach* — breath/wind/spirit, undivided — became Greek *pneuma*, which by then had already been split between the physical (air) and the metaphysical (spirit). The original unity was lost in translation — literally.

The history of the word 'spirit' is the history of progressive abstraction. And each step of abstraction was a step away from what the word originally captured: an invisible, unlocalized, causally effective process that the observer can feel within himself.

* * *

Here is what I ask you to hold in mind as you read the rest of this book.

The ancient speakers of Hebrew, Greek, Sanskrit, Chinese, and Slavic were not making a philosophical error when they used one word for breath, wind, and spirit. They were making a precise observation: there exist processes that are invisible, not localized in any particular substrate, causally effective through their manifestations, and continuous with the observer. And they named these processes with a word that preserved all these properties — because they had not yet learned to break the observation into pieces and file them in separate departments.

Modern thinking cut this observation apart. 'Breathing' went to physiology. 'Wind' went to meteorology. 'Spirit' went first to theology, then to the rubbish heap. We divided the territory into three kingdoms, declared them independent, and forgot they were once one.

This book is, among other things, an attempt to put them back together.

Not by returning to pre-scientific animism. Not by declaring that spirits are real in the way stones and electrons are real. But by asking a question that, once you hear it, is surprisingly hard to dismiss:

What if the ancient vocabulary — one word for breath, wind, and spirit — was tracking a real feature of reality that our modern specialized vocabularies have lost the ability to see?

And a second question, more personal, which I will be answering for the rest of the book:

What if a psychiatrist — a person professionally trained to distinguish insight from delusion — found evidence that the ancient vocabulary was right?

* * *

I need to tell you something about myself, because without it the book won't work.

I am a psychiatrist. My professional life consists of sitting across from people whose minds have done something unexpected — collapsed, expanded, restructured, broken — and trying to determine what happened and what to do about it. I was trained, at length and at considerable expense, to tell the difference between a mind working unusually and a mind working incorrectly. This distinction — between unusual and incorrect — is the most important tool in the profession, and the most poorly understood.

This training comes with a specific occupational hazard: you start seeing pathology everywhere. You watch a friend become passionately involved in a new spiritual practice, and a part of your brain whispers: prodrome? thought disorder? grandiosity? You read Castaneda describing conversations with a coyote, and instead of literary wonder you feel a diagnostic reflex: olfactory hallucinations, circumstantial thought, possible dissociative features. This reflex is not snobbery. It is professional muscle, built from years of sitting with people for whom 'unusual experience' was the first symptom of psychotic disintegration. The reflex protects. But it also blinds.

I first read Castaneda in residency. I didn't believe him. Of course not. I had been trained to distrust precisely the kind of experience he described: altered states, unusual perception, encounters with entities absent from the DSM. I filed his books under 'interesting ethnographic fiction' and moved on.

Except — when I now reread the opening paragraphs of this chapter, I catch myself feeling uneasy. Because Castaneda has a concept I did not and could not have known then: death as an adviser. Don Juan tells Carlos — as he always does, casually, without warning, in the middle of a walk through the desert: 'Death is our eternal companion. It is always to our left, at an arm's length. Death is the only wise adviser that we have. Whenever you feel that everything is going wrong and you're about to be annihilated, turn to your death and ask if that is so. Your death will tell you that you're wrong, that nothing really matters outside its touch. It will tell you: *I haven't touched you yet.*'

That November evening after the first death in my presence — when I walked home and suddenly noticed my own breathing — is retrospectively a textbook piece of death's counsel. But I could not have recognized it. I hadn't read Castaneda. When I eventually did — I didn't believe. When I retrospectively 'recognized' it — my professional alarm went off: ideas of reference, overvalued ideation, apophenia. The trap: without the framework, you miss the signal. With the framework, the recognition looks like pathology. Decide it's pathology and stop — and you may have discarded a real observation out of false caution. Instrument and object coincide; there is no external vantage point.

Do you see the structure? I can't win. If the coincidence is real, I'll miss it because I lack the framework. If the framework arrives and I see the coincidence retrospectively, it resembles pathology. If I decide it's pathology and withdraw, I may be discarding a genuine observation due to a false alarm. And if I decide it's real and proceed — I need to explain how I know it's not a delusion, using the very mind that might be producing the delusion. Instrument and object coincide. There is no external point from which to check.

But that's for later. For now, I was just walking home and breathing.

* * *

The observations stayed. Not the stories about Don Juan, not the desert landscapes, not the pyrotechnics of the nagual. Something quieter. Structural observations. Descriptions of how attention works. The mechanics of how a person interacts with a process larger than themselves and not quite visible. These observations, stripped of their mystical dressing, kept coinciding — with disturbing precision — with things I was seeing in clinical practice.

And every time I noticed a coincidence, the alarm sounded. Because I knew — professionally, by the book — what it looks like when someone starts finding 'confirmations' of their system everywhere. Apophenia. The experience of perceiving meaningful patterns in random data. A core feature of psychotic and pre-psychotic states. I was looking at what I saw — and simultaneously looking at myself looking, and checking: is this a pattern, or is this the beginning of one?

This tension drives the entire book. Each chapter is a double investigation: here is what I found, and here is the differential diagnosis I was forced to run on myself.

* * *

One more thing before we move on.

The pattern described in this chapter — one word for breath, wind, and spirit, independently arising in unconnected languages and civilizations — has a name in cognitive science of religion. It is an example of what Pascal Boyer calls a minimally counterintuitive concept: a concept that violates exactly one basic ontological expectation while respecting all others, making it simultaneously surprising and cognitively cheap to process. Spirits — invisible agents that act in the physical world — violate the expectation of visibility while preserving the full intuitive framework of agency: intentions, desires, goals, reactions. This is why the concept is 'sticky' — why it survives across cultures and centuries. It sits in the cognitive sweet spot between the banal and the incomprehensible.

Boyer's model explains why the concept is sticky — why it persists. But it does not explain why the concept is convergent — why independent populations arrived at the same conceptual identification. For that, you need a cognitive trigger: the unique phenomenology of breathing as an invisible, causally effective, universal process perfectly calibrated to activate the agent detection device. I described this trigger earlier in this chapter.

But here is the question that Boyer doesn't ask — and that I will be asking for the rest of this book:

What if the concept is not only sticky and convergent, but *accurate*?

What if there really is something in the world — invisible, unlocalized, causally effective, substrate-independent, and continuous with the observer — for the description of which the word 'spirit' was originally designed? Not an agent, not a ghost, not a soul in the Platonic sense. A pattern. An immanent pattern.

What if the ancients saw it and named it with the most precise word they had?

What if Plato broke that word, and we have been working with the fragments ever since?

The next chapter is about how I found the first fragment. It was in a book by Carlos Castaneda, and it was called an *inorganic being*. I didn't believe it. I still don't — not in the sense Castaneda intended. But the description worked. And that's when the trouble started.

Chapter 2. Inorganic Beings

Castaneda. First encounter with an explanation that shouldn't work but does.

Castaneda was handed to me by my future dissertation supervisor — at the time just a senior colleague I was learning from during residency. He said: read this, it'll be useful to you as a psychiatrist. You'll get patients who've read it, and you need to know the source material to tell the difference — is it a Castaneda-themed delusion, or did the person simply read too much and now thinks in those terms, or — and here's the hard part — did they actually *experience* something, and Castaneda turned out to be the only language that could describe it? Your job is to sort out which is which, and for that you need to know the material.

There was no official reading list — Castaneda wasn't part of any curriculum or recommended syllabus. It was the personal recommendation of one man. And I found it rough going — like a dental nerve under a trainee's drill: you know it's necessary, but the process is not one you recall with tenderness. There was no romantic reading with a worn paperback. I read a text file off the tiny screen of my phone — during the day on the *marshrutka* (a cramped minivan that passes for public transit in post-Soviet cities), squeezed between shifts, if I managed to get a seat and stay awake. At night, on my side job — I worked as a security guard in an empty building, and during the dead hours, when there was nothing to do but make sure nobody broke in, I sat in the silence of an empty hallway and read Castaneda off the palm of my hand. Mexican desert, Don Juan, peyote, some kind of 'warriors' — and outside the window, November, streetlights, and a shift at the ward in the morning.

* * *

I should probably say how my life was organized during that period, because otherwise it's unclear where everything that follows has its roots — and why I noticed what I noticed.

Besides residency and the night guard shifts, I worked as a videographer — shooting weddings, shooting nightclub parties, drawing animations and ad clips to order. This is not counting personal life and family, which also existed and also demanded attention. It sounds like a mess, and from the outside it was a mess. From the inside it was a centrifuge: you spin so fast that you stop noticing individual objects and start noticing the forces that hold them in orbit.

And here's the thing about side jobs. When you've shot your fifteenth wedding, something fairly strange happens: you stop seeing weddings and start seeing patterns. All weddings are the same — not in the details, in the structure. There's a set of elements that latch onto each other in a specific order, and the order doesn't depend on the people, the venue, the budget, or the country. The father of the bride always makes a specific face at a specific moment. The best man always goes slightly off-script in the same way. The first dance always has the same emotional arc. You notice this in the editing room — when you're cutting footage, stripping away the particular and searching for what works as a story. *Pattern* — I didn't know that word then, not in the sense I'll use it later in this book. But I was seeing one.

What matters — and what I didn't realize then but realize now as I write — is that it was in those side jobs, not in textbooks, not in lectures, that I first encountered the fact that a pattern is not an abstraction but a thing that works. You see it, and it gives you predictive power. A practical, measurable, mundane superpower: you know what will happen next, because you've seen the structure, and the structure repeats regardless of the material it's built from.

* * *

In parallel — first year of residency — I started attending courses in Neuro-Linguistic Programming. Most of what they taught struck me as what it is: repackaged behavioral therapy with a marketing gloss. But two things hooked me, and they hooked me for reasons I wouldn't understand until years later.

The first was the name Gregory Bateson, which surfaced occasionally in footnotes, without elaboration, as one of NLP's intellectual ancestors. Every time his ideas slipped through — fragments, quotes without context — something inside me said: *this person sees something the others don't*. I couldn't articulate what. Just a feeling — the kind you learn to trust much later and distrust at the time. I didn't read Bateson then. I reached him only when working on my dissertation, and the first book I read was *Angels Fear* — the posthumous one, 1987, co-written with his daughter Mary Catherine. But his name stuck. Filed under: come back to this.

The second thing that hooked me at NLP — and which I did not connect to Castaneda, or Bateson, or anything else I'm writing about now — was working with submodalities. I think it was called the Swish Pattern, a Bandler technique, though I may be confusing names — it's been years. The idea was this: you take a problem — a habit, a fear, something you want to change — and you represent it as a mental image. Not as a concept — as an *image* with specific sensory properties: size, brightness, distance, color, texture. Then you change those properties. Make the image smaller. Dim it. Push it away. And watch what happens to the problem.

At the time I thought: can you really change something real in yourself through a made-up thing? Turns out — you can, and it's not mysticism; there is literature showing that working with mental imagery through modification of its sensory characteristics affects emotional states. But that wasn't what struck me. What struck me was the operation itself. You take something — a problem, a state, a process — and you package it into an object. You give it sensory properties. It becomes a thing you can manipulate. A thing with quasi-agency — because the moment it has visual properties and a location in space, your brain starts treating it as something that can be addressed, moved, changed. Not the problem — the image. But through the image — the problem.

Now replace 'image with sensory properties' with 'spirit.' The operation is the same. Package a process into an agent, give it properties, interact with the properties — and the process changes. The only difference is the label.

I didn't ask this question at the time. It caught up with me later.

* * *

But back to Castaneda. In *The Art of Dreaming* — I didn't remember the exact chapter, because I wasn't reading very attentively — Don Juan describes beings he calls inorganic, and this word lodged in my mind because it felt strangely precise. Like someone had classified a ghost according to industrial standards.

The description went like this: beings without a biological body, without cells, without metabolism, without DNA — but possessing awareness and what Castaneda called an 'energy body.' They inhabit — and here, I remember, I looked up from the screen into the dark corridor — parallel dimensions of awareness, and can be contacted through specific practices, primarily through controlled dreaming. Contact with them can be useful but is dangerous, because the nature of these beings is fundamentally different from ours, and interaction can lead to either transformation or destruction of the practitioner.

I turned off the screen. Sat in the dark. And thought what any psychiatrist would think: well, here we are. Spirits.

* * *

I need to explain why this reflex is not snobbery or arrogance, because without this the rest of the book won't work.

When you spend years, every day, talking to people whose reality is coming apart — and psychiatric training, the very acquisition of phenomenological skill, is essentially training to see patterns: you're taught to describe a patient's experience without interpreting it, to note formal properties rather than content, to distinguish the structural from the incidental — you develop a filter. And the filter, over time, sharpens to a point where the word 'spirit' triggers a diagnostic cascade before the sentence is finished. Not because you're closed-minded. Because you've been trained by hundreds of cases in which 'spirits' meant *trouble*.

And the filter is correct — that's the thing. A person describing contact with invisible beings, in the majority of cases, needs help, not dream-practice development. I'd seen enough first psychotic episodes — voices that command, forces that control, beings that watch — to know: this is territory where it's better to over-diagnose. My future supervisor gave me Castaneda precisely for this — so I'd know where a patient's narrative was coming from, so I could distinguish cultural context from symptom. I learned. The Castaneda file sat somewhere in my phone's memory. Professionally useful, personally uninteresting.

Except that was not the end of it.

* * *

Several years passed. I finished residency, started working independently, and then something happened that I couldn't explain to myself for a long time.

A patient came to me — intact judgment, appropriate behavior, normal tests, a person no colleague of mine would hospitalize — and described an experience he could not name, and suffered precisely because he could not name it. Something was *present*. Not a voice — he was emphatic about this — not a hallucination, not a thought. A presence. Something that had no body but had qualities: density, temperature, something resembling intention. He could not see it. He could not describe it without sounding crazy, which was why he'd come to a psychiatrist rather than a priest or a psychic.

Standard psychiatry offers two answers: subclinical psychotic production or a dissociative mechanism. Both correct, both letting you fill in the chart and move on. But neither explains why this person — who had never read Castaneda, never attended an NLP course, never meditated a day in his life — described the experience in terms that converged, with disturbing precision, on what Don Juan told Carlos about inorganic beings. Same structure. Same properties. Same impossibility of naming.

* * *

By that time Castaneda had long since become 'Karlusha' among us — that's what we called him: me, my supervisor, other postgrads who'd also read and discussed him. *Karlusha* was our insider shorthand — a way to distance ourselves and simultaneously admit that we'd been hooked. You don't give someone a nickname if they haven't gotten to you. We weren't a fan club — we were a group of doctors who'd noticed that Karlusha's stuff works somehow, and couldn't explain how.

Here's what hooked us — and what is, incidentally, one of the most underappreciated things about those books: Don Juan was constantly laughing at Carlos. Not cruelly — more like at a child who keeps trying to understand a joke and can't, because he's looking in the wrong direction. Carlos would arrive with questions — serious, academic, anthropological — and Don Juan would howl. Sometimes literally roll with laughter. Because Carlos saw the world through one interface — rational, Western, academic — and was trying to cram into that interface something that didn't fit. Don Juan saw through two, and he found it hilarious to watch someone trying to describe an elephant through a keyhole. And — worse still — precisely because of this tunnel vision, Carlos was easily manipulated and didn't notice: if you see the world through only one filter, anyone who sees through two can steer you.

And one more concept that struck me as nonsense at the time and later turned out to be functional: the *ally*. Don Juan described inorganic beings not as enemies and not as friends — as allies. An entity you can work with if you know how. Dangerous not because it wishes you harm, but because its nature is incompatible with yours — yet if you learn to interact with it, it becomes a tool. I thought: right, the ally, the spirit helper, classic shamanic plot, nothing new. And then, in clinical practice, I encountered people who described their experiences in exactly those terms — and the word 'ally' fit better than anything the DSM offered.

* * *

I returned to the file not because I believed, but because the description *fit* — and when I reread it, with different eyes, after several years of clinical work, one word stopped me. *Inorganic.*

In everyday usage, 'inorganic' means 'not living' — rocks, salt, the periodic table. Castaneda seems to use it backwards: beings that are alive but have no organic body. Mystical nonsense at first glance. But if you approach the word the way my old etymology-obsessed Latin professor would — break it down — it opens up entirely differently. *In-organic:* literally, without organs. Without organization at the level of physical structures. Existing as a process, not a thing. Lacking substrate.

In philosophy of mind there is a precise term for this — *substrate independence* — and here I need to detour briefly into theory, because without it what follows won't make sense. Hilary Putnam in 1967 formulated the idea of multiple realizability: the same mental state can be realized on different physical substrates — neurons, silicon, whatever — what matters is not the material but the pattern of organization. Jerry Fodor in 1974 took it further: the regularities of the special sciences — psychology, economics, biology — cannot be reduced to physics precisely because the same pattern can sit on entirely different physical bases. Dennett in 1991 pushed it to the limit: consciousness is not a thing in the brain but a pattern the brain realizes.

And Bateson — the same one whose name at the NLP courses left me with the feeling that *this person knows something*. When I finally reached him — already working on my dissertation — it turned out he had formulated it earlier and more precisely than anyone: 'the pattern which connects' is not a property of any particular thing. It exists in *relationships* between things. Not in the crab. Not in the lobster. In the relationship between them.

A note, because Bateson is a figure easy to misread. He lived at Esalen — the California epicenter of the seventies New Age movement — and people who read him saw a mystic, because they wanted to see a mystic. His family was atheist, his science was rigorous, his cybernetics was mathematics. But he had the gift — or the curse — of seeing connections where others saw boundaries, and this made him look like a mystic to those who couldn't follow the logic. His final book, Angels Fear, co-written posthumously with his daughter, is titled with a phrase from Pope: 'For fools rush in where angels fear to tread.' It is a book about the sacred — written by an atheist, in the language of cybernetics. That is who Bateson was.

So: Bateson writes in the language of cybernetics, Putnam in the language of analytic philosophy, Fodor in the language of philosophy of science, and Castaneda in the language of beings — calling them inorganic. Four people, four languages, one structure. And the word 'inorganic' turned out to be precise — not metaphorically, but technically.

* * *

And here is what kept me up at night.

Castaneda's description — 'a being without an organic body, attracted by attention, capable of leading to transformation or breakdown' — worked as a predictive model. It predicted the structure of my patients' experiences — the same patients standard psychiatry could only describe as 'subclinical psychotic production.' Castaneda predicted the quality of attention that distinguished a working experience from a pathological one. He predicted that the form of interaction — whether the practitioner related to the process as an agent or as an abstraction — would determine the outcome. He predicted three possible outcomes: functional use, transformation, or breakdown.

A book by a questionable anthropologist about a Mexican shaman was predicting clinical phenomenology more precisely than the diagnostic manual compiled by hundreds of specialists over decades.

Try this on for size. 'A substrate-independent recursive pattern of cognitive organization, attracted by sustained non-directed attention' — the brain slides past, nothing hooks. Now: 'a bodiless being that feeds on your attention and can either help you or destroy you.' *Same content.* Different interface. The second version engages the emotional and social circuits that the first version bypasses entirely. Don Juan's language isn't less precise — it's more efficient, because it plugs into the full computational architecture of the human brain. But that's the next chapter.

* * *

A debt of honesty: there is a counterargument I cannot parry. Not yet. It was formulated by Jaegwon Kim, and it is brutal in its simplicity: if a pattern is not tied to a specific physical substrate, where does its causal power come from? What exactly produces the effects? Where is the stone that breaks the window — if the stone is a 'substrate-independent pattern'?

I don't have an answer. It will arrive in Chapter 5. For now, I want you to feel the tension, because I felt it every day: if what I'm observing is real — what is its ontological status? And if I can't answer — do I have the right to keep observing?

* * *

Around this period, my dissertation supervisor — the same person who'd once handed me Castaneda — said: 'Kirill, stop. I have concerns about you.'

He wasn't diagnosing me. He was doing what a senior colleague is obligated to do when the junior one has wandered somewhere worrying. But I knew what the formulation meant — 'close associates express concern while the subject's conviction remains intact' — that's a checkbox, I'd ticked it a hundred times on other people's forms. Along with 'subject reports insights inaccessible to others,' 'subject finds confirmations across unrelated domains,' and 'subject is unable to verify because the instrument and the object of observation coincide.'

And there were enough checkmarks: a growing number of 'confirmations' across different fields, emotional involvement, the feeling of 'special knowledge,' worried associates, and the crown jewel — the impossibility of external verification, because the instrument and the object are the same. By the book, this is a pre-psychotic prodrome. By the book — and I had written that book on other people.

* * *

I didn't stop — not because I dismissed his concern, but because I found the only criterion that permitted me to continue responsibly: does the model predict things I'm not looking for?

Apophenia works differently — you search for a pattern and find it everywhere, because the brain obligingly bends the data to fit the template. But if a model predicts things you weren't searching for, things that surprise you, things that are inconvenient, things you'd prefer not to find — this is a different kind of hit. It doesn't prove the model is correct. But it distinguishes the model from delusion: delusion always confirms. A working model sometimes disconfirms — and sometimes frightens.

And Castaneda's model predicted the thing that frightened me most: that the researcher cannot remain neutral. That studying these patterns will change the one who studies them. That instrument and object will inevitably coincide — not because of a methodological error, but because that is how the phenomenon works. I found this prediction confirmed on myself. And that confirmation — unlike all the others — could not be explained by apophenia.

* * *

Here is what I have at the end of Chapter 2: there are processes — I don't yet know what to call them properly, and I'll be circling toward that for several more chapters — that are invisible, not tied to a substrate, causally effective (or at least look that way), and describable in two languages: the language of science and the language of beings. The language of beings works better — and this fact, which should not be a fact, demands explanation.

And the question that determined everything I did next:

Why does description in the language of beings work better than description in the language of science?

Not because spirits are real, and not because science is wrong, but because — as I will argue in the next chapter — these are two different languages for the same thing, and one of them plugs into computational resources the other cannot access.

The next chapter is called 'Two Languages.' It is about why scientists, when they are really thinking, ask: 'what does the cell *want*?' — and why that question is not a metaphor.

Chapter 3. Two Languages

Why the same object can be described by a formula and by a spirit — and why the spirit is sometimes more precise.

In 2020, two biologists — Itai Yanai and Martin Lercher — published a paper with a title that made me drop everything I was doing and stay up reading until three in the morning. It was called 'The Night Language of Science.' The idea was simple and devastating at the same time: scientists use two languages. During the day — the strict, formal, publishable language of formulas, p-values, and mechanisms. At night — the loose, anthropomorphic, unpublishable language of intentions, desires, and agents.

During the day, a scientist writes: 'Cell proliferation is regulated by a cascade of signaling molecules, including...' — followed by half a page of abbreviations. At night, the same scientist tells a colleague over a beer: 'The cell wants to divide, but p53 won't let it.' And here's the thing — what Yanai and Lercher documented — the hypothesis, the unexpected idea, the creative leap almost always happens in the night language. Discovery lives in the night language. Publication lives in the day language. And the two are not allowed to meet in polite company.

I read this and realized they had described exactly what I'd stumbled on in the previous chapter, only from the other side. I'd found that describing a patient's experience in the language of 'beings' predicted better than the language of the diagnostic manual. Yanai and Lercher found the same thing in biology: the language of agents — 'what does the cell want?' — generates hypotheses that the language of mechanisms cannot reach.

* * *

Here I need to step back and say something that may be the most important thing in this book — something I could have said earlier but didn't, because I didn't know how. Now I do, and this, really, is what made it worth reading Castaneda, sitting in a dark corridor, and questioning my own sanity.

There is perceiving a process. And there is describing a process. These are different things, and they must not be confused.

Perception is when you are inside the process and aware of it from within, without a mediator. This is rare, difficult, and risky — Chapter 4 is about this, and there is plenty there to be afraid of. Not now. Now — description. Because everything we normally work with — formulas, poems, diagrams, spirits, metaphors, curses, prayers — all of it is *description*. Maps. Not the territory. Maps of different scales, different projections, different resolutions — but all maps.

If all descriptions are maps, then the question becomes: which maps are better? And the answer, which took me years to reach but is obvious once spoken aloud: it depends on what you're planning to do.

* * *

A detour — brief but important, because I have personal experience here, not just book knowledge.

I speak several languages. And here is what you notice if you stop and reflect: what you think — the actual thinking — happens in none of them. You just think. Thought exists before language — like pressure in a pipe before you turn the faucet. It's not Russian and not English, not Hebrew and not any other language you happen to speak. It simply *is* — formless, prelinguistic, a kind of cognitive plasma looking for a mold.

And when the faucet opens — when you start packaging the thought into words — you discover that different languages handle the same thought differently. There are things you can say quickly and precisely in Russian that take a three-clause detour in English. And vice versa — thoughts that settle into three English words and require an entire paragraph in Russian. A bilingual person doesn't think in two languages. They think in mentalese and speak in whichever language fits the shape of the thought at that moment.

This is not just introspection. Jerry Fodor — the same one who appeared in the previous chapter — proposed the 'language of thought' hypothesis in 1975: thinking occurs not in English or Swahili but in mentalese, a prelinguistic code that doesn't match any natural language. Steven Pinker made this accessible in *The Language Instinct*: language is a window onto thought, not the room itself. And Dan Slobin in 1996 formulated the concept that is key for this book: *thinking for speaking*

— language influences thought not in general but at the moment of packaging. When you need to say something, you begin organizing thought to fit the requirements of the language you're speaking in.

Lera Boroditsky showed that this goes further than grammar. Mandarin speakers organize time vertically — up is earlier, down is later. English speakers organize it horizontally — left to right. Same time. Different spatial packaging. The interface shapes what you see.

Do you see where this leads? If natural languages are interfaces for prelinguistic thought, and each interface handles some things better than others, then the language of formulas, the language of metaphors, the language of poetry, and the language of spirits are also interfaces. Just at a different level. Not for packaging thought into words, but for packaging *processes that don't fit into thought at all* — into a format the brain can work with. A bilingual person switches languages depending on what needs to be said. A person with two interfaces switches depending on what needs to be *done*.

* * *

Let's test this on a live example. Take a situation everyone knows: you've been building a plan for months — a project, a relationship, a career move, doesn't matter — and at some point you realize it's collapsing. Not gradually degrading but actively collapsing — fast, irreversibly, and there's nothing you can do. Describe this situation in five languages. Watch what each one does.

The language of formula: 'The negative deviation of the actual trajectory from the planned trajectory exceeded the threshold of the system's adaptive resources, resulting in a cascading failure of interdependent strategic elements.' Precise? Yes. Useful? If you're writing a paper on systems theory — sure. If you're living through this right now — the sentence slides off the brain like water off glass. No next step. No one to talk to. Dead language.

The language of metaphor: 'It all collapsed like a house of cards.' Better — you see the image, feel the fragility, understand the irreversibility. But the image is static: the house of cards doesn't look at you, doesn't want anything. You're watching from the outside.

The language of poetry: 'Standing in the wreckage of what I thought was solid, watching the wind carry the blueprints away.' Beautiful. Moving. But you're still a spectator — an emotionally engaged one, but a spectator.

The language of the body, unfiltered: 'The whole thing's fucked.' Compact, instantly understood, produces a visceral reaction — but generates no next step. A statement of fact, nothing more. In the Russian original this sentence uses *mat* — the profanity register I mentioned in the preface — and it hits harder than any English equivalent, because Russian profanity is a complete linguistic system with its own grammar, not just individual swear words. But the function is the same: acknowledge the disaster, generate no action.

Now — the language of the agent: 'Something came and destroyed my plan. It's stronger than me, and I need to reckon with it.' Feel the difference? This description does three things none of the others do. First, it creates an interlocutor — there is *someone* to deal with. Second, it creates a *relationship* — I and it, and between us there is a dynamic: it's stronger, but I can reckon with it. Third, it generates a *next step* — reckoning. Not acceptance, not observation, not description — action. Of all five languages, only the agent produces all three simultaneously: an interlocutor, a relationship, and a next step.

So if you need precision — take the formula. If you need to transmit meaning across distance — take the metaphor. If you need beauty — take poetry. If you need to swear — you have every right. But if you need to interact with what's happening — not just know about it, but do something with it — take the agent. Because the agent is the only description that plugs into the full computational architecture of the brain.

* * *

Why specifically an agent? Why not a formula, a diagram, a poetic image? I couldn't explain this to myself for a long time, until I came across several studies that, taken together, paint an embarrassingly clear picture.

Daniel Dennett — a man difficult to accuse of sympathy for mysticism — described in the 1970s what he called the *intentional stance*: a cognitive strategy in which you describe a system as if it has intentions, desires, and beliefs. Not because it actually has them — but because this description predicts the system's behavior more efficiently than any alternative. You can describe a chess computer through circuits, or you can say 'it wants to protect its queen.' The second description gives you more predictions per unit of cognitive effort.

Pascal Boyer — cognitive anthropologist, one of those authors I devoured — showed why this works, and his answer is elegant to the point of teeth-grinding: agent-concepts generate more inferences per unit of cognitive cost than any other type of description. Simply put: the brain, when it hears 'entity with intentions,' automatically generates a cascade of predictions — what it wants, what it fears, how it will react to resistance, what it will do next. No other conceptual format provides this density of prediction for so little effort.

And then — Deborah Kelemen, whose work finally demolished my confidence that science and 'mysticism' stand on opposite banks of an uncrossable river. Kelemen showed that even professional physicists — people trained for years to think in terms of mechanisms and equations — under cognitive load revert to teleological explanations. 'The river flows to reach the sea.' 'Bacteria evolve to develop resistance.' They know this is wrong. They would never write it in a paper. But under pressure, when the analytical channel is overloaded, the agent channel kicks in — because it's faster, cheaper, and more natural.

Elisa Järnefelt went further and showed that even committed atheists — people who consciously reject the existence of gods and spirits — retain automatic, pre-reflective, faster-than-conscious agent-detecting intuitions. At the level of fast cognition, everyone is an animist. Including you. Including me. Including Richard

Dawkins, though he would likely deny it.

* * *

Here's the picture.

We have two computational circuits in our heads. This is not a metaphor or a simplification — it follows from brain architecture, and if you need references: dual-process models of information processing are described by Kahneman, Evans, Stanovich and West, Lieberman, and despite all the differences between these models, they all converge on one thing: there is fast, automatic, emotionally loaded processing, and there is slow, controlled, analytical processing. These are not two 'modes' of one processor — they are, at least partially, *different neural substrates* that can work simultaneously, and which sometimes conflict.

A formula, a diagram, an equation — these are descriptions that work through one circuit: analytical, slow, sequential. You read a formula, and your brain computes it step by step, with effort, with resource expenditure. A powerful tool, but expensive: slow, energy-intensive, and limited in capacity — you can hold about four elements in working memory simultaneously, not forty.

An agent is a description that engages both circuits. When you describe a process as a being with intentions, the analytical circuit processes the content — what kind of being, what properties, what it does — while the emotional-social circuit simultaneously processes the *relationship*: is it dangerous? can I trust it? what does it want from me? This is not a metaphor — these are measurable neural processes: the amygdala, the medial prefrontal cortex, the temporoparietal junction — the entire Theory of Mind network lights up when you interact with an agent. Even a fictional one. Even an imagined one.

When a shaman fears a spirit, he is not deluded and not hallucinating. He is using a computational resource unavailable to the physicist writing an equation. Fear of the spirit is a way of processing information about a process. Trust in the spirit — also. Rage at the spirit — also. The entire repertoire of social cognition — millions of years of evolution, honed for interaction with *someone* — suddenly becomes available for working with processes that have no body.

And here I return to Castaneda, because he described exactly this — thirty years before Kahneman, in different language, but with the same structure. Don Juan distinguishes *first attention* and *second attention*. The first is our normal mode: the world described by words, organized by reason, held in place by what Don Juan calls the 'internal dialogue' — the continuous stream of mental speech by which we narrate reality to ourselves. The second attention is everything else: what opens when the internal dialogue falls silent, what is processed not by reason but by something else — something scientists would, thirty years later, call System 1, and Don Juan simply called 'the body.' Two attentions, two circuits, two languages. Day and night. Karlusha, for all his fabrications, described the architecture before neuroscience measured it.

* * *

And here I remembered the NLP courses and the submodalities — that technique I described in the previous chapter: take a problem, package it into an image, change the image's properties — and the problem changes. I didn't understand then why it worked. Now I do: packaging into an image engages the second circuit. The Swish Pattern is a stripped-down, secularized version of what shamans have been doing for millennia — translating a process into a format the full brain can work with.

This is why 'what does the cell want?' is a question that leads to discoveries, while 'what is the molecular mechanism of cell proliferation?' is a question that leads to the already known. The first question engages intuition, emotion, social cognition — and the brain starts generating hypotheses the analytical circuit alone would never produce. The second question engages only the analytical circuit — and gets a precise, verifiable, but unimaginative answer. Discovery lives in the night language. Publication lives in the day language. And science, by voluntarily abandoning the night language in the name of methodological purity, has been working with one hand tied behind its back.

* * *

Now we can return to Castaneda.

Castaneda — if you strip away the mystical wrapping, remove the desert, the peyote, and everything that makes him unacceptable to academia — does not describe spirits. He describes an interface. A way of interacting with bodyless processes by packaging them into the image of a being. Don Juan is not a mystic who 'sees spirits.' He is a user of the agent interface — someone who knows that describing a process as a being engages computational resources that describing it as a formula does not. He doesn't believe in spirits the way a parishioner believes in God. He *works* with them — as a tool, consciously, knowing the limitations.

Science versus magic is a false opposition. These are not two worldviews fighting over territory. They are two languages describing the same territory, each of which sees what the other cannot. The day language sees mechanism — and can verify it. The night language sees intention — and can generate hypotheses. The day language publishes. The night language discovers. And the person who speaks both — the person who can ask 'what does the cell want?' and then check the answer through experiment — sees more than the person who speaks only one.

* * *

Here, of course, a question arises, and I must answer it honestly before moving on: if both languages describe the same thing, why do we need the language of spirits at all? Why not just stick with formulas and add — well, good intuition?

The answer is that 'good intuition' is the second circuit working, just without a name. When a scientist says

'I have an intuition that the cell wants to divide,' she is already using agent language — she just doesn't admit it, because it's not acceptable in daytime. When a therapist says 'I feel something in the room,' she is detecting a pattern through the emotional-social circuit — and calling it 'clinical intuition' rather than 'spirit' is a matter of vocabulary, not of cognitive operation.

And here is the difference between Castaneda and the biologist who asks 'what does the cell want?' over a beer. The biologist is embarrassed by the night language and hides it. Castaneda — or rather, the tradition he describes — makes the night language the primary working tool. Not instead of the day language. Alongside it. Using both simultaneously, switching between them depending on what needs to be done.

* * *

I remember the moment it clicked — when all the pieces fell into place and I saw the whole. I was sitting at my desk with Yanai and Lercher, Boyer, Dennett, my Castaneda notes, and — for some reason — the rough cut of an ad I'd been editing the night before, all open simultaneously. And I suddenly understood: all of these are maps. Of the same territory. None of them is the territory itself. All of them are interfaces.

The difference lies in which computational resources the map engages. A formula engages the analytical circuit. A poetic image engages the sensory. A spirit engages everything: analytical, sensory, emotional, and social. This is why spirits 'work' — not because they are real in the sense that tables are real, but because they plug into the full bandwidth of the brain, while a formula uses one channel out of four.

But — and this matters — for the emotional and social channels to engage, you need *conviction*. You can't 'pretend' to fear a spirit. You can't 'sort of' believe. Without conviction, the pseudo-agent works like a formula — through one circuit. Conviction is not self-deception. It is a necessary condition for the interface to work — like current in a circuit. Without current, the circuit is just a diagram on paper.

* * *

I need to stop and say aloud what is probably already obvious, but which I am obligated to state explicitly, or I'll be dishonest: I am not claiming that spirits are real. I am not claiming that science is incomplete and needs mysticism. I am not claiming that Castaneda is a prophet and scientists are blind. I am not making any metaphysical claim whatsoever.

I am claiming that there exist processes which are real and causally effective but not tied to a specific substrate. That interacting with them requires a descriptive language. That several such languages exist — from formulas to spirits. That among them the language of agents is privileged because it engages two computational circuits instead of one. And that science, by voluntarily abandoning the night language in the name of methodological purity, lost access to a computational resource that other traditions — magical, religious, shamanic — preserved.

This is not an argument for magic and not an argument against science. It is an argument that bilingualism is better than monolingualism. That a person who speaks both languages sees more than a person who speaks one. And that Castaneda — with all his problems, fabrications, and questionable ethnography — provided a vocabulary that, at the descriptive level, is more precise than anything academic psychology has produced in the same territory.

* * *

And here is what genuinely angers me — not academically, but viscerally, in a way that makes me want to shout: people who use one language of description and don't realize it's only one of the available options. Who mistake their map for the territory and are prepared to go to war over its exclusivity.

On one side — people who foam at the mouth explaining that you're wrong, that spirits don't exist, that you're simply uneducated and lack sufficient knowledge to understand that everything is explicable by neurochemistry and information theory. On the other — people ready to devour you for having 'lost your humanity,' for becoming a cold, formula-obsessed drone who has forgotten the sacred and is destroying the mystery of life with his analysis.

And you haven't gone anywhere. You haven't left and you haven't arrived. You stand exactly where any person stands who has had enough curiosity, honesty, and stubbornness to look at both maps and notice they're drawn on the same paper. You are not a skeptic and not a believer. Not a scientist who has betrayed his calling, and not a mystic who has picked up a textbook. You are a person with two interfaces. You account for context and interlocutor. You choose the map that fits the task.

This, if you like, is the definition of freedom toward which this entire book moves. But we're not there yet. For now — we close Part I.

* * *

Here is what closes Part I — 'Rustling in the Bushes.'

Chapter 1: breath. The ancients noticed an invisible process and named it spirit. Plato broke that word, and we've been working with the fragments ever since. Chapter 2: inorganic beings. Castaneda described the same processes in the language of agents, and the description turned out to be more predictive than the scientific one. Chapter 3 — this one: why. Because the language of agents engages two computational circuits while the language of formulas engages only one.

Next comes Part II — 'The Mechanism.' It is about how this works from the inside. What exactly breaks in the brain when you try to comprehend a recursive process directly, without an agent-mediator. Why the ceiling exists. And what lies beyond it.

It is called 'The Ceiling.' And it begins with the time I tried to think what ought not to be thought.

Part II
The Mechanism

How what I discovered is built. A scientific argument lived as a detective story.

Chapter 4. The Ceiling

How what I discovered is built.

A scientific argument lived as a detective story.

Why the brain breaks on recursion — and why this is not a weakness but a feature.

Try this right now — it will take ten seconds, and you'll understand what this chapter is about faster than any explanation could deliver.

Pause for a moment and catch a thought that's in your head right now. Any thought — doesn't matter which. Just notice that you're thinking about something, and look at that thought as if from the outside.

Did it work? Of course it did — it's not hard. You noted a thought, maybe noticed it slipping away as you tried to look at it, and returned to reading. One level of reflection. Trivial.

Now switch: look not at the thought but at the process of observing the thought. Not what you're thinking — but how you're watching. Who is the one who was just looking at the thought?

Harder, but still manageable. Your head spins a little, there's a mirror-in-mirror sensation, but you can hold it.

Now — the third level. Look at the one who is looking at the one who is looking at the thought. Try to hold all three levels simultaneously: the thought, the observer of the thought, and the observer of the observer.

If you honestly tried, you know what happened: nothing. The thought collapsed. Instead of three levels — mush. And the only thing left is a slight dizziness and the feeling that you've run into something — like a glass wall, transparent, because you can see there's something on the other side, but you can't get through. This is what I call the ceiling.

That was about thinking. Now let's try the same thing with the body — and you'll see that the ceiling is not only cognitive.

Do the following. Contract your diaphragm — the flat muscle separating the thoracic and abdominal cavities — so that it descends and creates negative pressure in the pleural cavity. Simultaneously contract the external intercostal muscles so that the ribs rise and separate, increasing the volume of the thoracic cavity. Wait for the pressure gradient between the atmosphere and the alveoli to draw air through the nasopharynx, trachea, and bronchial tree into the lungs. Then relax the diaphragm, allow it to return to its dome position, simultaneously relax the intercostals — and let the elastic recoil of the lung tissue push the air back out.

Did it work?

Most likely not — or it worked with difficulty, with delays, clumsily, as if you were doing it for the first time in your life. But you are doing it right now. You were doing it while reading the previous paragraph. You were doing it while trying to hold three levels of reflection. You have been doing it every second of your life since birth. This is *breathing*. An ordinary respiratory act that you perform fifteen to twenty times per minute without thinking — but the moment it's broken into components and you're asked to execute them consciously, sequentially, by instruction, the process that was perfectly automatic becomes nearly impossible.

There's your ceiling in its purest form — only this time it's motor, not cognitive. You cannot breathe by managing breathing part by part, because the process is organized as a whole, and the attempt to decompose it into elements and reassemble it through conscious effort destroys precisely what makes it work. The instrument — conscious control — cannot contain the object — an automatic pattern. And notice: this is the same breathing that opened Chapter 1. The same invisible process that the ancients named spirit. They could not have described it as a sequence of muscle contractions — not because they were stupider than us. Because description by parts kills what they were observing.

Rutherford — the great experimentalist, whose biography I was reading in the same period as Bateson and Descartes — said: 'We are rather like children, who must take a watch to pieces to see how it works.' He meant it as a compliment to physicists. But when I read it, I thought the exact opposite: precisely — *you took it apart and it doesn't work*. A watch works while assembled. Disassembled, it's not a watch — it's a pile of parts. Breathing works while automatic. Broken into muscle contractions, it's not breathing — it's an instruction manual you can't breathe by. And a recursive pattern works while whole — disassembled into formal components, it ceases to be what you were trying to understand. The ceiling is the situation in which disassembly is impossible without losing what you're disassembling.

* * *

I first discovered the ceiling not in theory but literally — face-first — when trying to understand what exactly was happening with my attention during a long, focused session with a patient. I was sitting across from someone, listening, tracking their speech, noting patterns — and at some point caught myself tracking not just them but my own process of tracking. How I listen. What filters I engage. What I miss. This is normal supervisory reflection — it's taught, there's nothing mystical about it.

And then I tried to go further — to look at how I look at how I listen — and hit the ceiling. Not metaphorical, not philosophical. Cognitive. Thought refused to fold at the third level, as if working memory ran out mid-operation. The process I was trying to think about turned out to be larger than the instrument I was using to think about it. Which, when you think about it, makes sense: the instrument and the object are the same thing. The brain is trying to fit itself inside itself, and this fails for the same reason an eye cannot see itself without a mirror.

Except thought has no mirror.

* * *

What I found by face-planting has long been described in the literature — and described precisely, though I didn't know it at the time of the face-plant.

Fred Karlsson in 2007 conducted a systematic analysis of syntactic embedding in natural languages and found that in real — not laboratory, but live, conversational, everyday — speech, people virtually never use more than two to three levels of nested constructions. Not because the language won't allow it — grammatically you can nest sentences inside sentences to infinity — but because the brain refuses to process it. 'The cat that the dog that the boy that the grandmother called brought in scared ran away.' Grammatically impeccable. Cognitively — unusable. Three levels of nesting and you no longer know who did what to whom. Karlsson showed this is not peculiar to any one language — it is universal: all natural languages, on all continents, behave identically. Two to three levels — ceiling. Beyond that — mush.

Joanna Stiller and Robin Dunbar approached the ceiling from another direction the same year — through intentionality. They measured how many levels of someone else's consciousness a person can hold simultaneously. 'I know that you know that she knows that he thinks that I want...' — and found the limit at around five levels. Five is not the average — it's the upper bound, reached by few. Most people work comfortably at three or four. Shakespeare, by their calculations, built his dramaturgy at the fifth or sixth level of intentionality — which is precisely why his plays demand such cognitive effort and so reward it. But the sixth level is the ceiling of ceilings: beyond it, virtually no one can go.

And Ute Hahn and colleagues in 2022 provided the explanation that finally completed the picture for me: the issue is not that the brain is 'poorly designed' and can't handle recursion — it's that deep recursion is cognitively expensive and in most situations doesn't pay for itself. A resource-rational model: the brain isn't stupid, it economizes. Why spend computational resources on the fifth level of nesting if the fourth already gives a good enough result? The ceiling is not a malfunction. The ceiling is an *optimization*. The brain stops investing resources where the return drops below threshold.

And this last point is the key. Because it implies something I did not expect and that changed my view of everything this book is about.

* * *

The ceiling is not an error and not a weakness. The ceiling is an architectural property of a system that is trying to become aware of itself. A brain attempting to think its own process of thinking is a system whose instrument and object coincide. And there is a limit to how deeply a system can fold in on itself — just as there is a limit to how many times you can fold a sheet of paper in half. Not because the paper is bad — because that's how paper works.

Here the temptation — mine, and I suspect the reader's — is to invoke Gödel and Turing: the incompleteness theorem and the halting problem. Formal proofs that a system cannot fully describe itself. It's an attractive analogy, and I won't pretend it wasn't the first thing that came to mind. But I am consciously stepping away from it, and here's why: Gödel proved something about *formal systems*, not about brains. John Lucas and Roger Penrose attempted to transfer the Gödelian argument to human consciousness — since a formal system can't prove its own consistency, the brain must also have limits, and those limits are what makes consciousness special. Torkel Franzén and Paul Benacerraf disassembled this argument bolt by bolt and showed it doesn't work: limitations of formal systems do not entail limitations of biological brains. The transfer is illegitimate.

I accept their critique entirely. Gödel and Turing are not the foundation of my argument. They are the *cultural context*: the intellectual atmosphere of the twentieth century, in which the idea of principled limits on self-description became part of the air. But the foundation of my argument is not a theorem. The foundation is empirical. Karlsson. Stiller and Dunbar. Hahn. Real data on how real brains of real people behave when they try to fold in on themselves. And the data says: the ceiling exists, it is universal, and it is not overcome by greater intelligence or training.

It is built in.

* * *

And here — Metzinger. Thomas Metzinger, German philosopher of consciousness, the man who wrote a book with the magnificent title *Being No One* — and in it formulated the idea that became the final piece for my argument.

Metzinger speaks of the *transparent self-model*: the brain creates a model of itself, but this model is transparent — meaning the brain does not see it *as* a model. It sees through it — the way we see through glass without noticing the glass. You do not experience yourself as a model generated by the brain — you experience yourself as yourself, directly, without a mediator. This illusion of immediacy is necessary for normal functioning: if at every step you were aware that your 'I' is a construct generated in real time by neural activity, you couldn't tie your shoelaces, let alone lead a meaningful life.

The transparency of the self-model is the ceiling seen from the other side. The brain cannot fully become aware of its own process of awareness, because the instrument of awareness is awareness itself. You cannot step outside what you're looking with to look at what you're looking with. You can add levels of reflection — one, two, maybe three — but at some level the model stops being transparent, becomes opaque, and you hit the glass wall.

* * *

Now — the most important part of this chapter. What I built all of this for.

The ceiling exists. We don't overcome it. But we do something with it. And what we do with it determines — without exaggeration — the entire history of human culture, religion, madness, and freedom.

Castaneda — Karlusha, in our jargon — described the ceiling long before Karlsson and Stiller and Dunbar, only he named it differently. Don Juan says: the ceiling is the *internal dialogue*. The continuous stream of mental speech by which we narrate reality to ourselves — this is precisely what holds the world in its familiar form, precisely what won't let you past the ceiling. 'Stopping the internal dialogue is the key to the sorcerer's world,' says Don Juan, and if you translate from his language into neuroscience: cessation of verbal self-reference is the only way to weaken top-down priors enough for the ceiling to become permeable. What Carhart-Harris describes as REBUS under psychedelics, Don Juan describes as stopping the world — only without substances, through discipline of attention.

And one more concept without which this chapter is incomplete: *not-doing*. Don Juan explains to Carlos: 'doing' is everything your ordinary mind engages in — all the habitual operations by which you assemble and maintain the world. 'Not-doing' is *to stop doing what holds the ceiling in place*. Not inaction — an active operation of dismantling habitual cognitive pathways. In the language of neuroscience: the disruption of established attractors so the system can transition to a new configuration. In the language of this chapter: approaching the glass wall — not to smash it with a sledgehammer, but to discover that in certain places it is thinner than it seems.

There are three possibilities. Three states. Three ways to encounter the ceiling and three outcomes.

I need an analogy, and I'll take electricity — not because it's perfect (it isn't, and I'll say where it limps), but because it's good enough to make visible what without an analogy remains invisible.

* * *

First state: you interact with electricity through an interface. Light bulb. Socket. Voltmeter. You see manifestations, work with them, build models, predict behavior. You are not electricity — you are the one who uses it. This is all of science, all of mathematics, all of religion, all of magic, all of poetry. This is the second interface in its entirety — the language of formulas or the language of spirits, it doesn't matter, because in both cases you are on this side of the ceiling, working with descriptions. Descriptions can be precise or imprecise, useful or useless, beautiful or ugly — but they are all on this side. You are looking at electricity, not being it.

Second state: you have become electricity. The person you were is no longer there. There is no subject who 'perceives' the process — the boundary between perceiver and perceived has dissolved. This is not metaphor and not poetry. This is a state described by people who have undergone deep meditative or mystical experiences — and they describe it the same way, in every language and every culture: 'I ceased to exist,' 'I was everything,' 'boundaries disappeared.' You did not die — the one who usually watches dissolved. You are no longer observing the process — you are the process. Your manifestations are visible to others, but 'you' — the one who used to live inside — are not. The old cognitive architecture has been dismantled, and in its place something else has assembled, something that works differently and for which there is no term in the textbook, though in Castaneda's vocabulary there is one: 'man of knowledge.'

Third state: you tried to become electricity, and it cracked you. The old architecture dismantled, the new one didn't assemble. Between the second and third state there is no difference in intention — the difference is only in outcome. The same process, binary result: either it reassembled or it didn't. If yes — man of knowledge. If no — psychotic decompensation. Not metaphorical — literal, clinical, with hospitalization, antipsychotics, and slow, not always successful recovery. I have seen this. More than once. And this is precisely why it took me so long to accept what I was observing — because as a psychiatrist I saw primarily the third state, and the protective reflex said: everything that leads to the third state is dangerous. Stay away.

* * *

The reflex is correct. The third state is a catastrophe, and I have no intention of romanticizing it. Psychotic decompensation is not 'consciousness expansion that society doesn't understand.' It is a person whose reality has shattered, and putting it back together doesn't always work. I know what this looks like from inside the profession. I know the smell of a first-episode psychosis ward — a mix of institutional linoleum, bleach, and something else that has no name but that you recognize instantly because it's connected not to the place but to the state of the people in it. I don't need to be told why this is frightening.

But here is what took me years to understand: the third state is not an argument against the first and second. It is an argument that the ceiling exists for a reason. The ceiling is a protective structure. It won't let you into the territory where you might break. And when someone — through practice, substances, spontaneous shift, whatever — breaches the ceiling, the outcome depends not on intention and not on courage, but on whether the system can reassemble. If it can — second state. If it can't — third.

And in most cases — the first. You stand before the ceiling, press against the glass wall, and work with what's on this side. Descriptions, maps, interfaces — formulas and spirits. And this is not defeat. This is the normal, healthy, productive state. All of human culture was created in the first state — on this side of the ceiling.

* * *

Here I need to say where the electricity analogy limps. Electricity is substrate-bound: it has carriers — electrons, ions — that can be counted and measured. The processes I'm talking about — recursive, substrate-independent, not tied to any specific material — are substrate-bound only to the extent that a melody is: it exists, it's reproducible, it's recognizable, but you can't put it on a table. The metaphor limps at the crucial point.

But here is what I think matters: *any* metaphor for a recursive process will limp. Electricity limps on substrate-boundedness. Music would limp on linearity. Fire — on fuel consumption. Water — on flow without recursion. The very impossibility of finding a perfect metaphor is not a problem for my argument. It is a *demonstration* of my argument: a recursive process doesn't fit into any metaphor because metaphor is description through something else, and a recursive process is that which refers to itself. It cannot be described through something else without losing its essence. This is precisely why the only descriptive format that works at all is the agent — the being, the spirit: not because the spirit is accurate, but because the spirit is the only map that allows the brain to *interact* with what it cannot contain.

* * *

Here is the summary of this chapter, and at first glance it is disconcerting.

Recursive processes — those same invisible, substrate-independent patterns this book is about — do not cognitively fit inside the human brain whole. Not because we are stupid, but because the brain is part of what it's trying to think, and the instrument cannot contain the object when they coincide. The ceiling is real, universal, built into the architecture.

From this follow the two interfaces I've been discussing in previous chapters. The first: direct awareness — the attempt to breach the ceiling, with three possible outcomes. The second: description through an agent — working on this side of the ceiling but with both computational circuits engaged. And now that the mechanism is visible, we can ask the question that has remained unanswered until now: *what exactly is on the other side of the ceiling?* What is the process we cannot think whole but can interact with through interfaces?

The next chapter is an attempt at an answer. I will give this process a name: *immanent pattern*. And I will try to explain what I mean without resorting to formulas or spirits — and then show why you can't do without them anyway.

Chapter 5. The Immanent Pattern

The central term. What exactly we discover when we 'see' a recursive structure.

At the end of the previous chapter I promised to name what lies on the other side of the ceiling. It's time to deliver — and simultaneously to answer Kim's question, which has been hanging open since the end of Chapter 2 and hasn't let me sleep since.

The term I will use is *immanent pattern*. Not 'recursive algorithm,' though that's the closest formal equivalent. Not 'the pattern which connects,' though that's Bateson, and Bateson was right. Not 'spirit' and not 'inorganic being,' though these are working languages I use and do not renounce. Immanent pattern — because both words carry weight, and I need to unpack them before we go further.

* * *

Pattern is not a thing and not a property of a thing. A pattern is a structure of relationships that persists when everything else changes. When I was editing wedding videos and saw that the twentieth wedding was organized like the first, I was seeing a pattern. Not the specific faces, not the specific venue, not the specific cake

— but a structure independent of all of these, transferable from one wedding to another without loss. A pattern does not live in any particular wedding. It lives in the relationships between elements — which is why it is substrate-independent: it can be realized in different material, and it remains the same.

Bateson asked the question I consider the most precise question in all of twentieth-century philosophy — and asked it, as always, in a way that is impossible to dismiss: 'What pattern connects the crab to the lobster and the orchid to the primrose and all four of them to me? And me to you? And all the six of us to the amoeba in one direction and to the schizophrenic in another?' This is from *Mind and Nature*, 1979. *The pattern which connects is a metapattern*, he writes. A pattern of patterns. That which defines the vast generalization that, indeed, it is patterns which connect. I read this and realized Bateson had formulated what I'd been groping toward for years — only he fit it in a single question.

Immanent — and here the important part begins. In philosophy, 'immanent' is the opposite of 'transcendent.' The transcendent exists beyond the system, above it, outside it. A God who created the world and observes it from above is transcendent. A law of nature 'inscribed' somewhere in the structure of the universe is transcendent, at least in the popular understanding. The immanent is what exists *within* the system — inseparable from it, nonexistent without it. Breathing is immanent to the one who breathes: outside a breathing body, there is no breathing, only air. A melody is immanent to a performance: outside the sounding, there is no melody, only notes on paper — and notes are a different interface, a description, a map.

An immanent pattern, then, is a structure of relationships that does not exist outside the system in which it is realized, yet is not reducible to that system. It does not 'hover' above the substrate — it is *in* the substrate. But it is not *equal to* the substrate — because the same pattern can be realized in a different substrate, and it will be the same pattern. The way the same melody can sound on a violin and on a piano — it is in each of them, not above them, not without them, but not equal to either of them.

I realize this sounds like a philosophical puzzle — the kind with no solution. Wait. It's about to get concrete.

* * *

Let me show this with an example that causes no philosophical vertigo whatsoever — because everyone has encountered it and no one has ever considered it mysterious.

A traffic jam.

You're sitting in traffic, and ahead of you is a jam. It is real: you're burning time, burning gas, late for work, angry. It is causally effective: because of it you missed a meeting and the contract fell through. It has a location: right here, on this stretch of road, between this intersection and that one.

Now the question: what is it made of? Cars? No — the cars constantly change, some leave, others arrive; an hour later not a single car from when you entered the jam remains — but the jam is still there. The road? No — the road is the same when there's no jam. The drivers? No — the drivers have turned over three times. The jam is not a thing. The jam is a *pattern*: a stable configuration of relationships between elements that persists through the complete replacement of all elements. It is immanent to the flow — does not exist without the flow of cars, but is not equal to any particular car. It is substrate-independent in the sense that the same structure of jam can be reproduced on any road with sufficient traffic. And it is causally effective — you missed your meeting not because of a specific car but because of the pattern.

Nobody asks: 'Is the traffic jam real? Isn't it an epiphenomenon?' The jam is real in the most practical sense — it does things, it has consequences, it cannot be waved away. And yet it is not a thing, not an object, not a substance. It is an immanent pattern.

* * *

Another example — a little closer to what this book is about.

A stone arch. Each stone in the arch is just a stone: it has mass, shape, position. But the arch is not the sum of the stones. The arch is a configuration in which each stone holds *only because* the others hold it. Remove one — and what collapses is not it but the whole thing. The arch is an immanent pattern: it is in the stones, not above the stones, not without the stones, but not equal to the stones. From the same stones arranged differently, you get a wall — a different pattern, different properties, a different causal structure. Same stones. Different pattern.

What does the arch do? It *constrains*. It determines which configurations of stones are stable and which are not. It doesn't push the stones, doesn't pull them, doesn't do anything to them in the sense that gravity or glue 'does' something. It is a structural constraint: if the stones are arranged this way — it stands; if otherwise — it collapses. Gravity does the causal work. But which configuration gravity will sustain and which it will bring down — that is determined by the pattern.

Remember this word — constraint. It is about to close Kim's question.

And — since we're talking about constraints and about the invisible hitting the visible — Rutherford. I was reading his biography alongside Bateson and Descartes when I should have been writing my dissertation, and one sentence from his description of the gold foil experiment lodged in me so deeply I can still quote it from memory: 'It was almost as incredible as if you fired a fifteen-inch shell at a piece of tissue paper and it came back and hit you.' Rutherford was shooting alpha particles through gold foil, expecting them to pass straight through — and some bounced back. From this he understood that inside the atom there is something invisible, tiny, and incredibly dense — the nucleus. A structure that 'shouldn't have been there' revealed itself by striking the researcher. An immanent pattern works the same way: you fire through the model — and the pattern comes back and hits you. You can't see it, but you feel the recoil.

* * *

Back to Kim. At the end of Chapter 2 I said honestly: there is a counterargument I cannot parry. Causal exclusion: if every physical event has a sufficient physical cause, then the pattern — substrate-independent, multiply realizable — turns out to be an epiphenomenon, a shadow on the wall.

The answer I arrived at — and which, I should say, was not invented by me but by people far more competent in philosophy than a psychiatrist having an identity crisis — is this.

Kim is right about one thing: the pattern does not *act on* the substrate in the way a force acts. The arch doesn't push the stones. The jam doesn't push the cars. Natural selection — which Dennett, incidentally, called 'a substrate-neutral, mindless algorithm' — doesn't push genes. But Kim is wrong in his conclusion: from the fact that the pattern doesn't act as a force, it does not follow that it is causally inert. Because there is another type of causation — not force, but *constraint.*

A constraint is not what makes something happen. A constraint is what determines which things *can* happen and which cannot. A riverbed doesn't push water. But the riverbed determines where the water flows. Remove the riverbed — the water will still flow (gravity works), but it will flow entirely differently. The riverbed doesn't 'do' anything in the sense of applying force — but without it the landscape of flow is completely different.

In the same way, an immanent pattern does not 'act on' neurons. It determines which *trajectories* neural activity can follow and which it cannot. It is the riverbed, not the force. It is the topology of the space of possible states of the system. The causal work at the lower level is indeed done by physics — neurotransmitters, ion channels, action potentials. But which configurations of this physics are stable, which reproduce, which collapse — that is determined by the pattern. Just as the causal work in the arch is done by gravity, but whether the arch stands or lies in a heap of stones — that is determined by the configuration.

This, if you like, is Lagrangian causation instead of Newtonian: not 'what force acts on the object' but 'which trajectories are permitted and which are forbidden.' Walsh, Lewens, and Ariew formulated exactly this in 2002 with respect to natural selection: selection is not a force but a constraint on the space of possibilities. Kim is looking for a force — and not finding one. But he's looking for the wrong thing.

* * *

I remember the moment this reached me — not as a philosophical argument, but as a sensation, a physical 'aha,' after which the world looks slightly different.

I was thinking about natural selection — about how Dennett describes it, as 'an algorithm with no address, no substrate, no mind, but which shapes everything alive on the planet.' And suddenly I understood: nobody asks whether natural selection is real. Nobody asks whether it's an epiphenomenon. Nobody looks for the place where it lives or tries to weigh it. Everyone understands: selection is a constraint, and it is real in the sense that without it biological dynamics would be entirely different. Selection is an immanent pattern in its purest form: it is in the population, not above the population, not without the population, but not equal to any individual organism.

And then I thought: how is what I'm writing about different from this? Castaneda's inorganic beings, Bateson's patterns, the substrate-independent processes I encounter in clinical work — all of these are descriptions of something that works as a constraint, as a riverbed: does nothing in the sense of force, but determines which configurations are stable and which are not. And just as nobody doubts the reality of natural selection despite the fact that it can't be touched and has no address — there is equally no reason to doubt the reality of immanent patterns, even if they are invisible, unlocalized, and irreducible to substrate.

Kim is closed. Not refuted — closed: he was asking the right question but looking for the answer in the wrong category. He was looking for causal force, and what he needed was causal constraint.

* * *

Now — the three properties of the immanent pattern, gathered in one place, so there is something to push off from going forward.

Substrate independence. An immanent pattern is realized in a substrate, does not exist without a substrate, but is not identical to the substrate. The same melody on violin and piano. The same jam made of different cars. The same algorithm in neurons and silicon. This is not a 'freely floating' Platonic ideal — it is a structure that needs material but is not bound to any particular material. The way water needs a riverbed but can flow through different riverbeds.

Invisibility with causal effectiveness. A pattern cannot be seen, touched, weighed, or localized. But its effects are obvious and measurable: you were late because of the jam, the species survived because of selection, the arch stands because of the configuration. The pattern is causally effective — but not as a force, as a constraint. It determines the landscape of the possible; it does not push things across that landscape.

Coincidence of instrument and object. And this is the property from which everything written in the previous four chapters grows. When an observer tries to become aware of an immanent pattern, the instrument of awareness — their own cognition — is part of the very pattern they are trying to become aware of. This is not a defect of method and not a technical problem solvable by a more powerful instrument. It is a fundamental property: the observer is inside the observed. This is precisely why the ceiling exists. This is precisely why direct awareness of recursion leads to the three states described in the previous chapter. And this is precisely why interfaces are needed — formulas or spirits — to work with what cannot be contained whole.

* * *

In Castaneda there is an image for this — for all three properties simultaneously — that initially seems absurd and then turns out to be disturbingly precise: the *Eagle*. In *The Eagle's Gift*, Don Juan describes a force that generates awareness and reabsorbs it — not a being in the literal sense, but something to which the seers gave the name 'Eagle' because in seeing it resembled something infinitely vast, dark, with something like wings. This, if you strip the visual wrapper, is a description of the ultimate constraint — the system of which all observers, all patterns, all interfaces are part. Substrate-independent, invisible, causally effective, including the observer as part of itself. Three properties of the immanent pattern — in one image, in the language of an agent. And — strange as it sounds — Bateson's 'metapattern, pattern of patterns' is the same Eagle, only in the language of cybernetics.

* * *

Now we can step back and see how it all assembles.

Chapter 1: the ancients noticed an invisible, unlocalized, causally effective process and named it spirit. Chapter 2: Castaneda described such processes in the language of beings — and the description proved more predictive than the scientific one. Chapter 3: why — because the language of agents engages two computational circuits while the language of formulas engages only one. Chapter 4: the ceiling — an architectural limitation of a brain trying to become aware of a process of which it is part. And Chapter 5 — this one: *immanent pattern* — the term for what the ancients called spirit, Castaneda called an inorganic being, Bateson called the pattern which connects, and Dennett called a substrate-neutral algorithm. A structure of relationships irreducible to substrate, causally effective through constraint rather than force, and inevitably including the observer as part of itself.

The same thing — in four languages. And each language sees a piece the others miss.

One question remains — and it's technical, but without it the picture is incomplete: if immanent patterns are real, if they don't fit inside the brain whole, if interfaces are needed to work with them — what does this look like at the level of architecture? Not philosophically, but concretely: how does a system — any system, not just the brain — process recursion if recursion doesn't fit inside it whole? Are there examples beyond the human brain?

There are. Twenty thousand of them. They're called neural networks, and they process recursion *exactly the way we do*: through interfaces, not directly. That's the next chapter.

Chapter 6. Interfaces

Why machines confirm the thesis. Neural networks as the third illustration.

I need to explain something about my postgraduate years, because without it this chapter has no roots — and because I promised to be honest, and honesty here requires an admission that is not entirely flattering.

My dissertation was on self-stigmatization in patients with first-episode schizophrenia. Goffman, stigma, xenophobia as a survival foundation — a topic you can chew on indefinitely, one that will be forever dissertation-worthy because stigma is bedrock, it doesn't run out, and the academic council will always approve it. I chose it strategically. Not because it was my passion — because it was a ticket. In our system you cannot speak boldly from the podium until you have the degree — without it, people look at you *sideways*. The dissertation was the credential. The price of admission. 'Self-stigmatization in schizophrenia' — the council approves. 'Immanent patterns and pseudo-agent interfaces' — the council does not approve. Not yet.

Because what actually fired me up didn't fit any existing format. I was drawn to computer science, physics, epistemology, linguistics, data science, neural networks, and Castaneda — simultaneously, in one head — and I was convinced that between all of these you could build one large interface for psychiatry. That psychiatry is still in the Middle Ages: we diagnose by clinical impression and treat by trial and error, cycling through medications like keys on a ring. And what I wanted to build — the bridge between computational methods and clinical practice — required a credential. The dissertation was the credential.

So, in parallel with the dissertation — because I am constitutionally incapable of doing one thing at a time — I was reading Bateson, Descartes, and Rutherford's biography, enrolled in a master's program in computer science to learn data science and neural networks, and my master's project was a neural network that classified drawings by patients with schizophrenia versus healthy controls. After I moved to Israel I continued this work with a local population from local clinics.

Now — what happened when I got inside the neural network and saw how it's built.

* * *

A neural network does not think. This needs saying immediately, to prevent illusions: I am not claiming that machines possess consciousness, experience recursion, or 'see' immanent patterns. A neural network is a computational system that processes data. But precisely for this reason it is interesting for my argument — because it processes data without a second circuit, without emotions, without social cognition, without agents. It has one channel: numbers. And on this single channel — here is what the research shows.

Weiss, Goldberg, and Yahav in 2018 formally proved that standard recurrent neural networks (RNNs) cannot model recursive languages — the kind where you need to track nested structures, match brackets, maintain hierarchical dependencies. Michael Hahn in 2020 showed that even theoretically more powerful architectures — transformers, the foundation of modern large language models — have fundamental limitations on processing recursive dependencies. And Merrill and Sabharwal in 2023 demonstrated that the computational expressiveness of transformers, despite all their impressive capabilities, has a formal ceiling: there exist recursive structures they *cannot* process in principle.

And then Delétang and colleagues in 2023 ran what was at that point the most massive study: twenty thousand nine hundred and ten models, tested on tasks requiring the processing of recursive structures. Twenty thousand. And the result was uniform: no architecture processed recursion directly. All of them — every single one of the twenty thousand — built interface stacks.

* * *

An interface stack is, to put it simply, a chain of translations. A neural network doesn't take text and 'understand' it. It takes text and passes it through a sequence of transformations, each of which translates data from one format to another: tokenization (splitting text into pieces), embedding (converting pieces into numerical vectors), positional encoding (adding information about where each piece sits in the sequence), attention (determining which pieces are relevant to which), and only then — processing. Each layer is an interface: a translation from something the system can't work with directly into something it can.

When the interface fails, the model breaks. Not 'thinks incorrectly' — breaks in the literal sense: starts generating nonsense, loses coherence, hallucinates. Does this remind you of anything? It reminds me of the third state from Chapter 4: the old architecture has stopped coping, the new one hasn't assembled, and in the gap — mush. The neural network's 'psychosis' is structurally identical to the human third state. Not metaphorically — *formally* identical: loss of coherent output when the recursive depth of the task exceeds the capacity of the interface.

* * *

Here I need to tell you about the second job — the one that seemingly has no connection to this book but that gave me an understanding no book could provide.

After moving to Israel, alongside my postdoc, I got a job as a backend developer at an energy company. My task was maintaining payment acquisition services — all the payments people made for electricity. Sounds as far from Castaneda and immanent patterns as it gets. But here's what happened.

One day I was sitting there debugging yet another mapping between the payment gateway's format and our internal API, and I caught myself thinking: *I am doing exactly the same thing I think about in all my free time.*

Two formats that don't understand each other. Between them — an interface, a translation layer that converts one into the other without losing the essential information. The payment gateway speaks one language. Our system speaks another. Between them — a mapping. Remove the mapping — chaos. The wrong mapping — errors, lost transactions, money gone. The correct mapping — invisible, works silently, nobody notices it until it breaks.

A neural network processes text through an interface stack — tokenization, embedding, attention. A payment system processes a transaction through an interface stack — protocol, gateway, mapping, API. The human brain processes an immanent pattern through an interface stack — formula, metaphor, or spirit. Ancient man processed an invisible process through an interface — and called it ruach, pneuma, spirit. The principle is the same. No system works directly with what doesn't fit its format. There is always a translator.

Rutherford — whom I keep quoting because he formulated things with a precision any philosopher would envy — said: 'We've got no money, so we've got to think.' He meant the laboratory budget: when you can't afford expensive instruments, you have to invent. But for me the phrase sounds different: when there is no ready-made tool — *you build an interface from what you have*. Rutherford built from sealing wax and string — and discovered the atomic nucleus. The ancients built from rhythms and personifications — and created working cognitive technologies. I built a mapping between payment systems — and realized I was doing the same thing a shaman does. Everyone ran out of money. Everyone started thinking.

* * *

Here a counterargument arises that must be met head-on, because it is the most obvious and, at first glance, the most convincing.

The counterargument: neural networks process recursion without agency. Without spirits, without beings, without pseudo-agents. They simply compute. Therefore, the agent is not necessary; therefore, the thesis about the privileged status of the language of spirits doesn't hold; therefore, the whole construction collapses.

The answer: yes, neural networks process recursion without agency. And this *confirms* my thesis rather than refuting it. Here's why.

A neural network has no emotional-social circuit. It has one computational channel — numerical. It works only through the 'day language': transformations, matrices, activation functions. And it works — but with limitations. It cannot process deep recursion. It breaks on nesting that the human brain handles (albeit also with difficulty). The neural network demonstrates what happens with *only one computational channel*: you get less, and you break sooner.

A human working only through the 'day language' — only through formulas, only through the analytical circuit — behaves like a neural network: copes, but with limitations, and breaks on what doesn't fit the format. A human who has engaged the 'night language' — agents, spirits, pseudo-agents — has plugged in the second circuit and thus can process what a single circuit cannot. Not because spirits are real — but because the second circuit is a real computational resource that the first circuit does not duplicate.

* * *

Christiansen and Chater, as early as 1999 — long before the era of large language models — formulated the idea that proved key for this book: human language and neural networks are subject to the same computational constraints, because both are sequential processing of information under limited memory. And Hahn and Futrell in 2022 showed the punchline: the error patterns of the human brain and neural networks on recursive structures are statistically indistinguishable.

Think about what that means. Twenty thousand neural networks, silicon, possessing neither consciousness nor culture nor fear of death — and the human brain, biological, with all its history, emotions, and dreams — break on recursion *identically*. The same nesting levels. The same error types. The same collapse points. This is not a property of silicon and not a property of neurons. This is a property of the task. *Any* system processing recursion — regardless of what it's made of — hits the same ceiling.

And both systems — silicon and biological — cope in the same way: they build interfaces. Translate what doesn't fit into a format that does. The neural network translates text into embeddings. The scientist translates a pattern into a formula. The shaman translates a pattern into a spirit. I translated a payment format into an API. Same operation. Different substrates.

* * *

I want to pause and say something personal here, because without it the link between the energy company, Bateson, and Castaneda remains opaque.

The thing is, I'm wired to see everything at once. Not sequentially — simultaneously. When I was reading Bateson, I couldn't not think about neural networks. When I was debugging payment system mappings, I couldn't not think about the fact that the interface between a banking protocol and our internal API works on the same principle as the interface between the immanent pattern and the human brain. When I was working with a patient, I couldn't not see that the patient's attempt to describe their experience is the same kind of translation — from an internal format the system can't output directly into a format the interlocutor can process.

I know that from the outside this looks like chaos. A person simultaneously reading Descartes, Rutherford's biography, and TensorFlow documentation while working as a psychiatrist, a postdoc, and a backend developer — from the outside this looks like scatter, like an inability to focus, like ADHD that I don't diagnose because it works. But from the inside it was never chaos. From the inside it was always the same thing: the same pattern

— the same immanent pattern — manifesting in different substrates. The neural network, the payment system, the patient, the shaman, the arch, the traffic jam — *these are the same thing.* Not metaphorically. Formally. They share a common architecture. And my 'scatter' was simply the ability to see this architecture through different windows simultaneously. A dubious ability that makes you lonely, but once you have it, you can't unsee it.

And here is what I want to add in hindsight: the dissertation — that strategic ticket — really did open the door. I went not into 'some psychiatry' for my postdoc but into a computational psychiatry lab, where I worked on precisely what I'd been heading toward — that same interface between computational methods and clinical practice that I'd dreamed of building while sitting in a corridor at night with Castaneda on my phone. The scatter no longer looks chaotic. It looks like a pattern I couldn't see while I was inside it.

* * *

Here is the summary of Chapter 6, and simultaneously the summary of all of Part II — 'The Mechanism.'

Immanent patterns — invisible, substrate-independent, causally effective through the constraint of relational structure — do not fit inside any information-processing system whole. Not inside the brain, not inside a neural network, not inside a formula, not inside a ritual. This is not the weakness of any particular system — it is a property of the task: recursion does not fit inside a finite system that is itself part of the recursive process.

From this follows the necessity of interfaces — and it is universal. Not 'humans need interfaces because humans are weak.' Rather: any system needs interfaces for working with recursion, because recursion does not fit inside any finite system whole. The neural network builds a stack of transformations. The brain builds a stack of descriptions. The shaman builds a spirit. The physicist builds a formula. The payment system builds a mapping. All of them are doing the same thing: translating what doesn't fit into a format that does.

And now — armed with this understanding — we can look at what humanity has been doing with immanent patterns over the course of millennia. Not as a history of delusions, not as an anthropological curiosity, but as engineering practice. As thousands of years of interface development — blind, without a blueprint, but ruthlessly effective.

Part III is called 'Technology.' It begins with a chapter called 'The Blast Furnace' — and it is about how lightning becomes steel, and how a cognitive byproduct becomes a working technology.

Part III
Technology

How it worked for millennia. Religion, ritual, altered states — not errors but engineering.

Chapter 7. The Blast Furnace

How it worked for millennia.

Religion, ritual, altered states — not errors but engineering.

Religion as cognitive technology. Why Boyer is right — and why it isn't enough.

Lightning strikes a tree. The tree catches fire. This is not technology — it is a byproduct of atmospheric physics.

A human picks up the burning branch and carries it to a cave. This is no longer quite a byproduct — but it is not yet technology either. It is the accidental exploitation of a physical phenomenon: fire existed before humans; humans simply learned to pick it up.

A thousand years pass. Humans have learned to start fire themselves — by friction, by flint, by whatever means. This is closer to technology, but still primitive: you're working with what nature gave you, you've just learned to reproduce the conditions.

Several more millennia pass. The forge. The kiln. The smelter. The blast furnace — a construction that allows you to control temperature with a precision no campfire can match and to produce materials that do not exist in nature: steel, glass, ceramics. Between the lightning that struck the tree and the blast furnace lie millennia of cultural work, and at no point was there an engineer with a blueprint. A blind process in which more effective practices displaced less effective ones, mistakes were weeded out, successful discoveries were transmitted.

A byproduct of atmospheric physics became technology not because someone 'designed' it. But because millennia of cultural evolution turned accidental fire into a controlled process.

This, I argue, is what happened to religion.

* * *

Pascal Boyer — one of those people whose books I read with a pencil in one hand and irritation in the other, because he is *right,* and that only makes it worse — in 2001 formulated the position that has since become mainstream in the cognitive science of religion: religion is a byproduct of cognitive mechanisms that evolved for other purposes. The Hyperactive Agency Detection Device (HADD, discussed in Chapter 1) evolved to spot predators — and religion accidentally exploits it. Theory of Mind evolved to predict the behavior of conspecifics — and religion accidentally uses it for communicating with gods. Coalition psychology evolved to maintain the group — and religion accidentally builds churches on it.

Scott Atran said roughly the same thing in different words in 2002: religion is a spandrel, a side effect of cognitive architecture that was not selected for by evolution and performs no adaptive function. It exists because the cognitive mechanisms that produce it are useful for other things.

Both are right. And I say this without irony and without qualification: Boyer and Atran are right. Religion really did arise as a byproduct. Agent detection really did evolve not so that people would talk to spirits. Theory of Mind really did evolve not so that people would pray. This is accidental exploitation of cognitive mechanisms — and if the story ended here, I would close this book on this page and go do something useful.

But the story doesn't end here. Because between a byproduct and a technology lie millennia of cultural evolution. And across those millennia, something happens that Boyer's framework does not cover.

* * *

Manvir Singh in 2018 published a paper on shamanism that is, in my view, one of the most underappreciated articles in the cognitive science of religion — and it says exactly what Boyer is missing. Singh showed: yes, shamanic practices *began* as accidental activation of cognitive mechanisms — rhythm, repetition, altered states, personification — but then cultural selection kicked in. Practices that 'worked' — that more reliably activated the right cognitive modes, produced more stable effects, more predictably induced certain states — were transmitted, modified, refined. Those that didn't work were forgotten. After thousands of years of this blind, engineerless, but ruthlessly effective process, what remained was not a random byproduct. It was a polished technology.

Harvey Whitehouse documented this from another angle in 2004: he identified two modes of religious practice that convergently arise in unconnected cultures around the world. The imagistic mode — rare, high-affect rituals that are remembered for life: initiations, ecstatic practices, conversion experiences. And the doctrinal mode — frequent, low-affect, repetitive rituals: daily prayer, weekly services, reading of texts. Both modes are not accidents. Both are optimized for different cognitive tasks: the imagistic for forming episodic memory and group bonds through shared experience, the doctrinal for forming semantic memory and transmitting knowledge across generations. Two tools — two types of tasks. This is not a byproduct. This is *engineering* — if engineerless.

* * *

And here is where it gets truly interesting — the part that stitches this chapter to the previous six.

Norton and Gino in 2014, then Hobson and colleagues in 2017, ran experiments that, in my view, change everything. They showed that rituals work — in a measurable, experimental sense: they reduce grief, lower anxiety, modulate the neural error response — and they work not through belief but through formal structure. That is: you don't need to believe in God for a ritual to have an effect. You don't need to belong to a tradition. You don't need to know the 'meaning' of the ritual. It is enough to perform a certain sequence of actions with certain formal properties — rhythm, repetition, structure — and the cognitive effect occurs.

Moreover: *arbitrary, freshly invented* rituals — with no cultural history, no connection to any tradition, devised by the experimenter an hour before the experiment — produce the same effect. Formal structure works *before* any tradition. Tradition is millennia of optimizing what works without it.

Do you see where this leads?

If the formal structure of a ritual works on its own — without belief, without tradition, without cultural context — then the ritual is an interface. Just like a formula in science, like an embedding in a neural network, like a mapping between payment systems. A way of translating what doesn't fit into the cognitive system directly into a format the system can work with. Rhythm, repetition, personification — these are not 'ornaments' and not 'vestiges of magical thinking.' They are formal properties of an interface, optimized by millennia of blind cultural evolution.

* * *

Now — the central question of this chapter, the one for which it was worth building the blast furnace analogy.

Why do rituals *converge*?

Why do cultures that never contacted each other — in Amazonia and Siberia, in West Africa and Polynesia — converge on the same formal properties in their ritual practices? Rhythm. Repetition. Personification — addressing an agent, a spirit, a being. Altered states — through dance, breathing, sound, fasting, substances.

Group synchronization. Narrative — a story told while the ritual is performed.

Boyer explains this through cognitive architecture: the brain is built the same way in all humans, so the byproducts are the same. This is correct but insufficient. Byproducts explain why the concept of a spirit arises everywhere — but they don't explain why the ritual practice of working with it converges on a specific format. Boyer explains why lightning strikes trees on every continent. He doesn't explain why furnaces on different continents are built similarly.

The answer is the same one that explains the convergence of furnaces: *the task is the same*. If you want to smelt iron, there is a limited number of ways to do it. Physics doesn't depend on culture: you need high temperature, controlled airflow, a specific chamber geometry. Cultures arrive at similar solutions not because they copy each other but because the task imposes constraints on the space of possible solutions. Constraint — as in Chapter 5.

Likewise: if the task is to provide a cognitive system with access to an immanent pattern through a pseudo-agent interface, the formal requirements for the interface are the same, because the cognitive architecture is the same. You need rhythm — because rhythm synchronizes neural oscillations. You need personification — because the agent engages both computational circuits. You need an altered state — because the normal processing mode is optimized for everyday life, not for interaction with what lies beyond the ceiling. Cultures arrive at the same ritual forms not because they exchange ideas but because they are building an interface to the same thing.

* * *

I remember when this thought first entered my head — and I remember recoiling from it, because it sounded like I was about to say something New Age, and I would sooner bite off my tongue than say something New Age.

But then I checked myself — with the same criterion that saved me at the end of Chapter 2: does the model predict what I'm not looking for? And the model predicted.

It predicted that rituals which have *lost* their formal properties — in which the rhythm has faded, personification has weakened, altered states are no longer induced — should stop working. Not 'stop seeming important' — stop producing measurable cognitive effects. And this is precisely what is observed in secularized, ritually impoverished forms of religious practice: when all that remains of a ritual is social convention — 'we go to church on Sundays because that's what you do' — the effect vanishes. It is not that belief departs — the interface stops working because the formal properties that made it an interface have been removed.

And it predicted the reverse: that *new* rituals, constructed with the right formal properties but without any cultural tradition, should work — and work from the first time. This is exactly what Norton and Gino showed: an invented-an-hour-ago ritual reduces grief. Not because anyone believes in it — but because its formal structure meets the requirements of the interface.

* * *

Now I can say what I could not say in Chapter 1, because then it would have been ungrounded, and now — after five chapters of argument — it stands on a foundation.

Religion is not an error and not a byproduct. More precisely: it *was* a byproduct — just as fire from lightning was a byproduct of atmospheric physics. But millennia of cultural evolution turned the byproduct into technology — just as millennia of working with fire turned the campfire into the blast furnace. Religious practices — rituals, prayers, meditations, ecstatic experiences — are cognitive technologies for accessing immanent patterns through the pseudo-agent interface. Not the only technologies — science is also a technology, working through a different interface — but the only ones that engage both computational circuits.

And when I say 'technology,' I don't mean it as a metaphor. I mean: a set of reproducible procedures with a predictable outcome, optimized through iteration and transmitted across generations. That is the definition of technology. The blast furnace meets it. The shamanic ritual meets it. Daily prayer meets it. If something looks like technology, works like technology, and is transmitted like technology — perhaps it's time to stop calling it superstition.

* * *

Costly signaling — the theory that religious rituals exist to demonstrate group commitment (Richard Sosis) — explains why rituals persist. But it doesn't explain why they take the form they do. If it were only about signaling, the ritual could be anything — hopping on one leg, wearing red hats, reading the multiplication table aloud every day. But rituals converge on specific formal properties — rhythm, personification, altered states — and costly signaling doesn't explain this. Cognitive function does: rituals have this form because it is the form of a working interface.

Robert McCauley proposed an analogy I find precise: vision is a byproduct of the evolution of photoreceptors. The telescope is a technology. Between them lie thousands of years of cultural work with optics. Nobody says that the telescope is 'a byproduct of photoreceptors.' Likewise, it is wrong to say that a shamanic ritual is 'a byproduct of agent detection.' The byproduct was the first spontaneous detection of an agent in the rustling of bushes. The ritual is the telescope.

And here I return to Castaneda once more, because he has one of the most unsettling — and simultaneously most technically precise — images in the entire series: the *flyers, voladores*. In *The Active Side of Infinity*, Don Juan describes beings that feed on human awareness — literally consuming the 'glowing coat of consciousness,' leaving a person with just enough to survive but not enough to be free. Sounds like paranoid delirium — and that is exactly how any psychiatrist, including me, would read it. But strip away the visual wrapper and translate into this book's language, and the flyers are a description of parasitic patterns: processes that exploit cognitive architecture, consuming attention and leaving a person functional but unfree. Social protocol, habit, automatism — everything I saw when my life flashed before my eyes under the sound of a siren in Israel. And the defense against flyers that Don Juan describes — 'inner discipline' — is, in essence, a description of calibrated attention: a technology that allows you to distinguish where your awareness is working for you and where it is feeding a pattern that lives off you.

* * *

And — closing on the personal — this is why what I saw in clinical practice worked. The patient who described a 'presence' — an invisible process behaving like an agent — was using an interface. Not because he was ill (though some were). Not because he was uneducated (some were more educated than me). But because the pseudo-agent interface — a technology honed over millennia — gave him access to something for which formal language had no words.

And this is why my dissertation supervisor was right when he asked me to read Castaneda: not so that I would learn to diagnose the over-read, but so that I would begin to see behind the 'mystical delusion' — a technology. Not always — sometimes it really is delusion, and an antipsychotic is necessary. But sometimes it is a working interface, and the psychiatrist's job is not to destroy it but to understand what it gives access to.

The next chapter is about that. It is called 'Haloperidol Won't Help,' and it is about something every psychiatrist knows but nobody says out loud: that there are things you cannot cure, because they are not a disease.

Chapter 8. Haloperidol Won't Help

Altered states through the eyes of a psychiatrist. Why it works, why it's dangerous, and how to tell transformation from breakdown.

I had a dream during residency — one of those dreams you don't forget, because they don't dissolve in the morning but instead grow sharper year by year, as if it's not you remembering the dream but the dream gradually remembering you.

I dreamed I was being admitted. Not anywhere — to the first-episode psychosis ward. The same one I walked past every day on the way to work, where I knew every nurse, every smell, every sound — the electronic lock, the click, the soft thud of the door closing behind you. Only this time it closed behind me not as doctor but as patient.

The doctor — I don't remember his face but I remember his voice, calm, routine, the voice I myself used with patients when delivering a decision they would not agree with — the doctor said: we'll start with haloperidol. Standard protocol.

And I, in the dream, answered with perfect calm: it won't help. There's no point of application.

The doctor looked at me with the expression I had seen on my own face in the mirror — when a patient said something that sounded reasonable but was part of a delusional system: a blend of attention and professional compassion. He nodded. Made a note in the chart. Haloperidol, intramuscular.

I woke with the feeling that the dream was more precise than waking life. Not in any mystical sense — in a technical one. Because the phrase 'there's no point of application' was not delusion but diagnosis. Haloperidol is an antipsychotic, a D2 dopamine receptor blocker. It works when the problem is excessive dopaminergic activity in the mesolimbic pathway: the classical model of productive psychotic symptomatology — voices, delusions, thought disorganization. But if what is happening to me is not excess dopamine but a different processing mode, a different interface, access to immanent patterns through a mechanism absent from the DSM — then haloperidol literally has no target. It does not treat what is not a disease. It silences a signal without considering that the signal may have contained information.

* * *

Only a psychiatrist can write this chapter. I say this not from professional snobbery — I say it because any other author at this juncture either romanticizes or demonizes, and both are dangerous.

The philosopher discussing altered states from an armchair romanticizes: 'consciousness expansion,' 'access to deeper layers of reality,' 'transcendence of the mundane.' Beautiful words that smell of an incense boutique and a complete absence of clinical experience. The philosopher has not seen a person in an altered state bang their head against a wall because they cannot stop the flood of perception. The philosopher has not held the hand of a mother being told that her twenty-year-old son, after a 'spiritual experience' at a retreat, is now in a ward and may or may not come back.

The mystic who romanticizes madness is even more dangerous: 'madness is misunderstood wisdom,' 'the shaman's illness is an initiation,' 'society suppresses those who see more than others.' Every time I hear this I remember specific faces — people whose lives were destroyed by a psychotic episode, and for whom romanticizing their suffering would not have helped one gram. Psychosis is not initiation. Psychosis is catastrophe. Sometimes reversible, sometimes not.

And the rationalist who demonizes any unusual experience — also wide of the mark: 'it's all brain chemistry, an illusion, a byproduct of neural noise.' This explains some cases — but not all. And it closes a door that perhaps should have been left ajar.

A psychiatrist — an honest one — stands on the boundary. He has seen catastrophe and knows what it smells like. He has seen transformation and knows it exists too. And his task is not to romanticize or demonize but to distinguish one from the other. A differential diagnosis. Not a metaphor — a professional obligation.

* * *

The three states I described in Chapter 4 as a theoretical construct — these are not an abstraction. They are what I have seen with my own eyes, in clinic, repeatedly.

And here I want to describe them not as a philosopher but as a physician, because the difference in language is a difference in precision, and precision here costs lives.

First state — working through the interface. The person practices — meditates, prays, performs a ritual, uses a psychotechnique — and returns. They can describe what they experienced. They can draw a line between 'ordinary' and 'unusual' states. They remember who they are, where they are, what day it is. Judgment is intact — they understand that the experience was an experience, not reality in the ordinary sense, though they feel it was significant. This is safe. This is normal. This is, in essence, what any therapist working with imagery, visualization, or guided fantasies does. The Swish Pattern from Chapter 2 is in this category. The interface works, the person uses it, the ceiling is in place.

Second state — transformation. The old cognitive architecture has been dismantled, and something different has assembled in its place. The person is different — not in the romantic sense of 'became better,' but structurally: they process information differently, see patterns they didn't see before, act from a different position. They may be disoriented during the process — sometimes severely, sometimes for a long time — but ultimately they cohere. Whole, but different. In Castaneda's vocabulary this is 'a man of knowledge.' In clinical terms — I don't know what to call it, because the textbook has no term. There is no diagnosis of 'successful cognitive transformation.' There is only 'remission' — but remission presupposes disease, and here there was no disease.

Third state — breakdown. Dismantling without reassembly. The old architecture has already been taken apart, the new one hasn't come together, and the person lives in the gap — or, more precisely, doesn't live but exists in a state that clinically is called psychotic decompensation. Shattered reality, lost boundaries, inability to tell inside from outside. This is not 'consciousness expansion.' This is catastrophe. The person is in a ward, nurses, electronic lock, haloperidol intramuscular.

Between the second and third state there is no difference in intention. No difference in sincerity. No difference in 'spiritual level' — whatever that means. The difference is only in outcome: did it reassemble or not. The same process, binary result.

* * *

Here I will say what is not customarily said, and what will probably displease colleagues on both sides of the barricade — both psychiatrists and practitioners of 'spiritual' techniques.

We *cannot* distinguish the second from the third in real time. Not 'we're bad at it' — we cannot. When a person is in the process of transformation — when the old architecture has already been dismantled and the new one has not yet assembled — they are clinically *indistinguishable* from a person in psychosis. Thought disorganization — present. Impaired judgment — present. Unusual content of experience — present. Close associates express concern — present. Every checkbox ticked — the same checkboxes I was counting on myself in Chapter 2.

The distinction can be made only retrospectively — by outcome. If it assembled — it was transformation. If it didn't — it was psychosis. But in the moment — when a decision must be made: give the antipsychotic or don't, hospitalize or don't — we have no instrument for the distinction. And we typically make the safe choice: hospitalize, medicate, silence the signal. Because the cost of the error 'didn't hospitalize, should have' is a human life. And the cost of the error 'hospitalized, shouldn't have' is an interrupted transformation that is impossible to prove.

It is precisely this asymmetry — the cost of error differs in the two directions — that creates the situation in which we systematically suppress the second state, mistaking it for the third. And I am not claiming we are wrong to do so: in the absence of an instrument for the distinction, the safe choice is the only ethical one. I am claiming we need the instrument. And that the first step toward building it is to acknowledge that the second state exists.

* * *

Now — drugs. Because no book that talks about altered states can avoid this topic, and I don't intend to.

Psychedelics — LSD, psilocybin, DMT, mescaline — in the model I'm building here, are not 'another path to the same place.' They are the *forcible dismantling of old cognitive architecture without any guarantee that a new one will assemble*. This is the key distinction, and it needs to be understood.

Robin Carhart-Harris and Karl Friston in 2019 proposed the REBUS model — Relaxed Beliefs Under Psychedelics — which describes what psychedelics actually do to the brain: they weaken top-down priors, the entrenched expectations through which the brain filters incoming information. Under normal conditions, these priors are a blessing: they let the brain avoid processing every stimulus from scratch, conserve resources, live in a predictable world. A psychedelic weakens them — and suddenly you're perceiving the world *without filters*, or with minimal filters, and it feels like 'consciousness expansion' because the volume of incoming information spikes.

But — and here begins a conversation romantics don't enjoy — weakening priors is not wisdom. Weakening priors is removing protection. What will fill the space left by the removed filters is not predetermined. Sometimes — a new, more accurate model of the world. Sometimes — chaos. Sometimes — a new configuration that is worse than the old one. A psychedelic opens the gates but does not control what walks through.

Letheby and Gerrans in 2017 described what happens at the subjective level: ego dissolution — the dissolution of 'I,' the sensation that boundaries between self and world have vanished. In my model this looks like this: the ceiling has been temporarily breached, and the person finds themselves in territory their usual cognitive architecture is not adapted to. If the architecture manages to restructure — the person comes back different, and sometimes 'different' is better. If it doesn't — the person comes back broken. If the restructuring begins but doesn't complete — the person lives in an intermediate state, neither here nor there, with weakened filters and unfinished assembly.

I have seen all three outcomes. Not in books — in clinic. And here is what I took from that experience: drugs are a sledgehammer where a watchmaker's tool is needed. They solve the access problem — the ceiling is breached, the gates are open — but they create the control problem: what happens next, you don't control. Castaneda's Don Juan, incidentally, says exactly the same thing: power plants are a tool for beginners, for those who need to be shown that something exists beyond the ceiling. But the real work requires other methods.

* * *

Here I need to tell you something autobiographical that bears directly on what I've just said.

When I started working as a psychiatrist, I told myself firmly: addiction medicine — never. Not my thing. Don't like it. Don't want it. But life decided to have a laugh, and during my postgraduate years it handed me a job — simultaneously in several rehabilitation centers for drug and alcohol addicts. And it was there — in rehab, among people who had been through everything I'd just described theoretically — that I got the chance to test whether I had actually understood anything by studying psychedelic experience, Castaneda, altered states, and all of this 'mysticism' from the outside.

Because — and this needs saying directly — I have never in my life taken psychedelics. Not on principle and not from fear — there simply was no desire. When you are already in a state that could be called 'heightened awareness' — reading Castaneda and Bateson and Descartes simultaneously, training neural networks, working as a psychiatrist — chemical expansion on top of all that doesn't attract. It alarms: if the normal mode is already running near capacity, why increase the risks?

But here I need a caveat, because without it any experienced clinician reading this book will say: 'So what? After five years of practice I also know what a patient will tell me — that's called clinical experience, not revelation.' And they would be entirely right. Any addiction specialist with years of experience predicts the structure of an addict's narrative — the phases, the typical insights, the characteristic self-deceptions. This is normal professional development. There is nothing unique about it, and I have no intention of passing off routine clinical competence as mystical vision.

The difference is this. My colleague in addiction medicine also sees the pattern. They predict in clinical language: 'now comes denial, then bargaining, then — if we're lucky — acceptance.' They work through one interface. What I discovered was that *the very same pattern* — the same sequence, the same structure, the same inflection points — is described by Castaneda in the language of beings and by Bateson in the language of cybernetics, and that all three descriptions — clinical, mystical, and cybernetic — are maps of the same territory. The uniqueness is not in seeing the pattern. The uniqueness is in seeing it from several interfaces simultaneously and understanding that it's the same one. My colleague predicts no worse than I do — but they don't know that Castaneda describes exactly the same thing. I do, and this knowledge changes not the prediction but the understanding of *why* the prediction works.

So — *I knew what they would tell me before they told me*. Not the content — the content differs for everyone, as the model predicts: the form of the interface depends on culture and biography. But the structure — the pattern of the experience, the sequence of phases, the type of insight, the character of what 'opens,' and the type of danger that lies in wait — is the same for everyone. I could predict the moment a person would say 'I understood something important but can't explain it.' Could predict when they'd describe the dissolution of boundaries. Could predict what kind of fear would arise and when.

My interlocutors in rehab — the patients — were convinced I'd also been a bit of a user. How else could I so precisely understand what they were talking about? I hadn't used. I simply knew the pattern — and could switch between interfaces freely enough to talk to them in their language and then switch back to clinical. This, in essence, is the bilingualism I wrote about in Chapter 3 — only lived, not described.

Granted, over time even this became predictable to the point of boredom — like weddings in the editing room. The pattern repeated, the faces changed, the structure remained. But I loved that work, and — as far as rumors reached me — the patients liked working with me too. Probably because I was one of the few psychiatrists who didn't write off their experiences as 'side effects of intoxication' but talked about them as real experience with structure. Didn't indulge, didn't romanticize — but didn't dismiss.

And incidentally — it was in rehab that I truly came to appreciate one of Don Juan's strangest recommendations, which until then had struck me as eccentricity: *be tasteless.* Don Juan explains to Carlos that a warrior must be 'tasteless' to the forces that feed on human awareness — not attract attention, not be a delectable morsel. In the context of rehab this acquired an entirely concrete meaning: when you work with people whose processes — addictions, compulsions, patterns of self-destruction — literally feed on the attention and energy of those around them, you need to be able to be present without becoming food. Not to emotionally engage to the point where the patient's process begins working through you. To maintain distance — not cold distance, but calibrated distance. To be tasteless is, in essence, a psychiatrist's professional hygiene — only described in the language of beings rather than the language of the textbook.

* * *

Here a counterargument will inevitably arise — and I can hear it coming, so I'll answer.

David Nichols, the neuropharmacologist, will say: *'The feeling of insight induced by a psychedelic is not equivalent to real insight — it is a side effect of serotonergic activation, and attributing epistemic value to it is an error.'* He is right that the feeling of insight ≠ the presence of insight. A patient in a manic state also feels they've understood everything — and usually hasn't understood a thing. But from the fact that the feeling of insight does not guarantee insight, it does not follow that there is no insight. The question is not in the subjective experience — the question is whether the 'information' obtained predicts anything new. Same criterion that saved me in Chapter 2.

Steven Katz will say: *'Mystical experience is culturally conditioned — a Buddhist sees emptiness, a Christian sees light, a shaman sees animals, and this proves the experience is constructed, not discovered.'* He is right that the form of experience is culturally conditioned — a Buddhist really does see emptiness, not angels. But in my model this is not a problem but a confirmation: culture determines the form of the interface — which pseudo-agent appears, what 'clothing' the pattern wears — but does not determine what the interface gives access to. The Buddhist's emptiness and the Christian's light are different maps of the same territory. Different interfaces, same immanent pattern.

Zoran Josipovic — a neuroscientist studying meditative states — in 2014 and 2019 described what he called 'nondual awareness': a state in which the boundary between subject and object of perception dissolves, but without loss of cognitive clarity. Not psychosis. Not hallucination. Not ego dissolution. Consciousness that knows itself as conscious without mediation by image or concept. In my model this is closest to the second state — transformation in which the ceiling has not been smashed by a sledgehammer but has become permeable, and the person can exist on both sides simultaneously without breaking.

* * *

And now — the most personal thing in this book.

I have walked this boundary. Not as an observer — as a participant. Periods of prolonged focused work, meditation, clinical practice in which every day you interact with people whose reality is configured differently — all of this brings you to the ceiling, and sometimes the ceiling thins, and you feel something showing through from the other side, and you need to decide: go further or stop.

And every time — every time without exception — when I felt this permeability, the same alarm went off that I've been writing about since Chapter 1. Because I know what the third state looks like. I know its smell — a mix of institutional linoleum, bleach, and something else that has no name but that you recognize instantly because it's connected not to the place but to the state of the people in it. I know what the eyes of a person whose reality has shattered look like. I have seen this dozens of times — and each of those times left a notch, and those notches form a reflex that says: careful. It's dangerous here. You could crack.

And here is what it means to be a psychiatrist on this boundary: you cannot romanticize, because you know the price. You cannot demonize, because you see that beyond the ceiling is not emptiness. And you cannot stop checking yourself — because instrument and object coincide, and the only guarantee that you're on the right side is that you keep checking.

The one who has stopped checking — either passed through transformation and no longer needs the check (second state), or lost the ability to check (third state). And distinguishing between them — only retrospectively.

* * *

My dream was precise in one thing: haloperidol won't help. Not because I've lost my mind. But because haloperidol is a tool for the third state, and I am in the first, with periodic approaches to the second. An antipsychotic working on D2 receptors literally has no target in what I'm describing. It's like treating a fever with antipyretics in someone who is hot because they're running: the temperature is elevated, but the cause is not infection — it's a process.

And here — the closing argument of this chapter — I want to return to what I said at the beginning: only a psychiatrist can write this chapter. Because only a psychiatrist can simultaneously know that haloperidol won't help and know that sometimes it is absolutely necessary. Not everyone who hears voices is a shaman. Not everyone who sees spirits is working with an interface. Sometimes it is disease, and disease requires treatment, and haloperidol saves lives — literally, not metaphorically. My task is not to abolish antipsychotics. My task is to show that in some cases what's needed is not an antipsychotic but a different instrument — and that for this one needs to be able to distinguish.

And to distinguish, one needs to acknowledge that there is something to distinguish. That the second state is not a variant of the third, not 'mild psychosis,' not 'subclinical psychotic production.' That it exists as a separate phenomenon. And that the first step toward working with it is to stop automatically treating it with something it doesn't respond to.

* * *

I know what my inner opponent is thinking right now — the one who has been sitting in my head since Chapter 1 and whom I cannot silence because he speaks in my own voice: *'How do you know you're in the first and not the third? How do you know this entire book isn't the product of a slowly building delusional system in which you can no longer tell insight from pathology?'*

I don't know. That's the whole point. Instrument and object coincide, and there is no external point from which to check. All I can do is keep doing what I've done from the start: check whether the model predicts something new. Check whether the data confirms. Listen to people who say 'stop,' and each time decide — is this calibration or is this noise. And acknowledge that certainty will never come.

This is probably the least comforting sentence in the book. But it is also the most honest. And, as I will show in the next chapter, this impossibility of stepping outside is not a problem. It is the foundation of the entire argument.

The next chapter is called 'The Observer Inside.' It is about why you cannot step beyond the system — and why precisely this makes knowledge gained from within not less but more reliable.

Chapter 9. The Observer Inside

Why you cannot step outside — and why this is not a problem but a foundation.

If you have read this far, a question should already have occurred to you — one I have deliberately postponed, and which, if not addressed now, will undermine everything said in the preceding eight chapters.

The question is this: *how can this book assert anything, if the author is inside the system he is describing?*

I claim that immanent patterns are real, substrate-independent, and causally effective. But I myself am part of these patterns. My cognition is part of what I describe. My brain, which is building the argument, is part of the system the argument is about. Where is the guarantee that the observation is not distorted by the observer's position? Where is the guarantee that I am not constructing a self-confirming system — delusional, if you like — that looks logical from inside but bears no relation to reality?

This is not a rhetorical question. It is the most serious epistemological objection possible, and if I don't answer it, the entire book is a pretty construction hanging in midair.

* * *

The objection is ancient. It is thousands of years old, and it is called the problem of self-refutation: if every assertion is conditioned by the observer's position, then the assertion 'every assertion is conditioned by the observer's position' is also conditioned, and therefore cannot claim truth. A snake biting its own tail. Any theory that claims the observer is inside the observed undermines its own foundation, because the theory itself is an observation from inside.

If you are my dissertation supervisor, or a reviewer, or simply a thinking reader — this is where you put the book on the shelf and say: logical error, self-refutation, no need to read further.

But don't rush. Because this problem has been solved — independently, from entirely different traditions — by at least eight thinkers, and their solutions, taken together, do not merely dissolve the objection but turn it into an argument.

* * *

Niklas Luhmann — the German sociologist who is nearly impossible to read but who, if you do push through his prose, rewards like few others — said: yes, the observer is inside. This is not a problem — it is a *condition of observation*. A system that observes itself cannot observe its own observation at the same moment — there is a 'blind spot,' a structure that makes observation possible but is itself invisible from within that observation. This is not a defect — it is an architectural property. The way an eye cannot see itself — but from this it does not follow that the eye does not see.

Humberto Maturana — the Chilean biologist, one of the creators of the concept of autopoiesis — approached it from the other side: the observer cannot step out of life to describe life, but can describe *themselves describing*. This is not complete self-description — there will always be a layer that is not captured — but neither is it total blindness. Partial transparency, which increases with each level of reflection — up to the ceiling discussed in Chapter 4.

Douglas Hofstadter — author of *Gödel, Escher, Bach*, the man who made self-reference a subject of pop culture — showed that *strange loops* — systems that, moving through a hierarchy of levels, return to the beginning — are neither an error nor a paradox. They are a source of new properties absent from any individual level. Self-consciousness, in Hofstadter's account, is a strange loop: a system that models itself modeling itself, and it is precisely this loop that generates what we call 'I.' Embeddedness is not a bug. Embeddedness is architecture.

And Descartes — whom I was reading alongside Bateson and Rutherford when I should have been writing my dissertation, and whom I adore for precisely one thing — did something that is usually overlooked when people quote 'cogito ergo sum.' His famous 'I think, therefore I am' is not merely the foundation of rationalism. It is the *first documented case in Western philosophy of instrument and object coinciding*. Descartes tried to doubt everything — absolutely everything, including the existence of the world, the body, other people, God, mathematics — and discovered that there is one thing that cannot be doubted: the very act of doubting. Because in order to doubt, there must be someone who doubts. The instrument — doubt — ran into the object — the doubter — and found they were the same thing. The observer inside the observed. Descartes essentially discovered the ceiling — only instead of acknowledging it, he tried to build all of metaphysics on top of it, and from this came a catastrophe four centuries long: dualism, the split between mind and body, and everything that followed. But the moment of discovery itself — brilliant.

Nāgārjuna — a Buddhist philosopher of the second century CE, and if it seems strange to you that he stands alongside Luhmann, Descartes, and Hofstadter, you haven't read Nāgārjuna. He formulated this two thousand years before everyone else: any assertion, including the assertion of the emptiness of assertions, is itself empty — and this does not refute it but *demonstrates* it. The assertion 'everything is conditioned' is itself conditioned, and this is not a paradox but direct confirmation of what it asserts. The snake is not biting its own tail — the snake is completing the circle.

Graham Priest — a contemporary logician, specialist in paraconsistent logic — formalized what Nāgārjuna described intuitively: there exist statements that are simultaneously true and false — *dialetheia* — and a logical system capable of handling them does not explode but functions. Self-refutation is not the end of the argument but an indication that classical bivalent logic does not cover the entire territory.

Gödel — whom I already discussed in Chapter 4 and

promised not to overuse: his incompleteness theorem says not 'the system is useless' but 'the system cannot be simultaneously complete and consistent.' From an embedded position one can obtain an enormous number of true statements — just not all of them. And that is a limitation, not a refutation.

And finally — a concept I did not find in any single author but rather assembled from several: *virtuous circularity*. Varela and his enactivist colleagues described the cognitive circle — the observer shapes the observation, which shapes the observer — not as a vicious circle but as a generative one. Each pass through the circle does not return to the same point but ascends to the next turn of the spiral. This is not running in place. This is ascent.

* * *

Eight thinkers, eight traditions, one thought: the embeddedness of the observer is not a defect and not a weakness. It is a condition of a particular type of knowledge that is impossible to obtain from outside.

And here — Michela Massimi, philosopher of science, and her concept of *perspectival realism*, which in my view closes the question.

Massimi says: objectivity is not a 'view from nowhere.' Objectivity is the convergence of multiple perspectives, each of which is limited, each conditioned by the observer's position, but which, taken together, yield a picture unavailable to any of them individually. The twelve decimal places in the measurement of the electron's mass were obtained not from an 'objective' position outside physics — but from the convergence of hundreds of experiments, each embedded in physical reality, conditioned by a specific instrument, a specific laboratory, a specific experimenter. And yet — twelve decimal places. From an embedded position.

Helen Longino — social epistemology — took this to its logical conclusion: knowledge is reliable not when obtained from a 'neutral' position (no such position exists) but when it has passed through a system of *social verification* — criticism, replication, peer review, challenge. Every participant in the verification is inside the system. None stands outside. But a multitude of embedded observers, each with their own blind spot, correct one another — and the result turns out more reliable than any single 'objective' observer could provide, were such a being to exist.

* * *

Now — application to everything this book is about.

If I could step outside and look at the immanent pattern from the side — objectively, from 'nowhere,' without involvement — I would not need interfaces. I would simply *see* it, the way I see a table or a chair, and describe it the way I describe a table or chair — unambiguously, unconditionally, without equivocation. But I cannot step outside. My brain is part of the system I am describing. My cognition is one of the immanent patterns I am trying to think. Instrument and object coincide.

Precisely for this reason two interfaces are needed. Not because we are weak. Not because we lack computational power. But because what we are trying to know includes us as part of itself, and the only way to work with it is either to dissolve into it (the first interface, direct awareness, with three states and all the attendant risks) or to project it onto a screen the brain can process (the second interface, formulas or spirits).

Embeddedness is not a weakness of the argument. Embeddedness is its foundation. The entire argument of this book rests on the observer being inside the observed. If the observer were outside — there would be no ceiling, no three states, no need for interfaces, Kim's problem would not arise, and there would be no need to write this book. The book exists because the observer is inside.

* * *

And finally — what I was thinking when my dissertation supervisor asked me to stop, when the inner opponent asked 'are you sure you're not in the third state?,' when every confirmation of the model triggered the apophenia alarm.

Knowledge obtained from inside a system is not less reliable than knowledge obtained 'from outside.' It is *of a different type*. It is perspectival — tied to a position. It is incomplete — there is a blind spot, and it cannot be eliminated. It requires verification by other observers — Longino, social epistemology — because a single embedded observation may be distorted, while the convergence of many is reliable.

But it has one property that knowledge 'from outside' does not and cannot have: it knows what it is like to be part of what it describes. The observer who watches a thunderstorm through a window knows one thing about the storm. The observer who stands in the rain knows another. Neither of these knowledges replaces the other. But to pretend that knowledge from the window is 'objective' while knowledge from the rain is 'subjective noise' — that is not scientific rigor. That is an arbitrary restriction of epistemology.

Bateson knew this — and, unlike Descartes, did not try to climb out. In the afterword to *Steps to an Ecology of Mind*, he wrote a sentence that is worth the entire book: 'The individual nexus of pathways which I call "me" is no longer so precious because that nexus is only part of a larger mind. The ideas which seemed to be me can also become immanent in you. May they survive — if true.' Here is an observer who knows he is inside — and does not panic but *works* from there. Does not try to exit. Does not lapse into relativism. Says: I am part of something larger, my knowledge is incomplete, but it can become part of your knowledge — if it is true. Perspectival realism, formulated thirty years before Massimi.

This book is written from the rain. I am not standing at the window describing immanent patterns from outside — I am inside them, I am part of what I am talking about, and my instrument coincides with my object. This makes my knowledge incomplete — yes. It makes it risky — yes, I have felt this since Chapter 1. But it also makes it a type of knowledge impossible to obtain from behind the window.

And the convergence — Massimi's perspectival realism — works. Castaneda describes from inside. Bateson describes from inside. Putnam describes from inside. My patients describe from inside. The neural network behaves as the model predicts. Rituals work as the model predicts. Four descriptions from four different positions — all embedded, all conditioned, all incomplete — and all converge. Twelve decimal places from the rain.

* * *

This closes Part III — 'Technology.' We have traveled from the blast furnace to the observer inside the system, and now we can finally ask the question the whole structure was built for: what do you do with all this? How do you live if you see two interfaces? How do you not go mad and not become a charlatan? How do you tell a precise observation from a hallucination? And — the most personal question of all — what does it cost?

Part IV is called 'Freedom.' It begins with a chapter called 'Calibration' — and it is about what distinguishes a mage from a paranoid. The answer: calibration. Only calibration.

Part IV
Freedom

How a psychiatrist became a mage. What it costs. And why it is happiness.

Chapter 10. Calibration

How a psychiatrist became a mage. What it costs. And why it is happiness.

Every interface can be used precisely or imprecisely. The difference is calibration.

Vision produces optical illusions. A stick dipped in water looks broken. The moon on the horizon looks enormous. Two line segments of equal length look different if you draw arrows on their ends pointing in different directions. The human visual system — one of the most sophisticated in the animal kingdom — systematically, reproducibly, predictably errs.

And nobody concludes from this that vision is useless. Nobody proposes abandoning eyes because they produce illusions. We do something else: we learn to distinguish when vision shows accurately and when it lies — and we introduce correction. Lenses, instruments, cross-checking with other senses. We don't abandon vision — we calibrate it.

Now replace 'vision' with 'agent detection' — and you have the problem around which this entire chapter is built.

* * *

Van Elk in 2013, and Willard and Norenzayan the same year, showed that hyperactive agent detection — the same HADD discussed since Chapter 1 — correlates with paranormal beliefs. People whose HADD runs at elevated RPMs see spirits more often, believe in ghosts more often, interpret random events as intentional more often. Pennycook and colleagues in 2012 showed the reverse: an analytical cognitive style — the habit of thinking slowly, checking intuition, not trusting first impressions — suppresses agent intuitions. The more analytical the person, the less often they see agents where there are none.

At first glance — an argument against everything this book is about: agent detection produces false beliefs, analytical thinking suppresses them, therefore we need more analytical thinking and fewer spirits. Close the book, go do science, stop muddling people's heads.

But there is a catch, and it lies in the word 'suppresses.'

* * *

Suppression is not calibration.

Suppression is when you kill the signal entirely. Analytical thinking, taken to its extreme, suppresses agent intuitions — all of them, indiscriminately. It does not distinguish false positives from true ones — it simply switches off the detector. This is effective against errors: you stop seeing spirits in the rustling of bushes, and that's good, because there are usually no spirits in the bushes. But you simultaneously lose access to the second computational circuit — to what the language of agents does better than the language of formulas. You stop seeing false patterns — but along with them you stop seeing real ones, those accessible only through the agent interface.

This is like treating a migraine by cutting off the head. Effective? Yes, the migraine no longer bothers you. But the price is disproportionate.

Calibration is different. Calibration is when you don't switch off the detector but *adjust its threshold*. You still see agents — because you need this interface, need access to the second computational circuit, need the ability to ask 'what does this thing want?' and receive an intuitive answer. But you learn to distinguish: this is a pattern, it is real, it predicts, you can work with it. And this is noise — a false positive, an artifact of the detector being too sensitive. A calibrated instrument does not go silent — it tells the truth more often than it lies, and you know when to trust it and when to double-check.

* * *

And now — the definition that is perhaps the most important in this book, because everything in the final chapter follows from it.

The difference between a paranoid and a mage is *calibration*. Only calibration. Nothing else.

The paranoid sees agents everywhere — and cannot stop. Their detector runs at maximum, threshold at zero, every rustle is a message, every coincidence a sign, every passing glance surveillance. They cannot distinguish signal from noise, because *everything* has become signal. An uncalibrated instrument.

The skeptic is the polar opposite: they have suppressed the detector to zero, switched off agent intuitions, live in a world where there are no spirits, no agents, no 'night language,' only formulas and p-values. They make no errors — in the sense that they have no false positives. But neither do they see — what is visible only through the agent interface. A suppressed instrument.

And the mage is a calibrated instrument. They see agents — because they need the interface. They know that some of what they see is noise. They can distinguish. And — this is key — they can stop and check. The paranoid doesn't stop. The skeptic doesn't start. The mage starts, stops, checks, and goes on.

* * *

And here my dissertation supervisor returns.

Every time he said 'stop, I have concerns about you,' he was not hindering me. He was not suppressing my detector. He was not trying to 'cure' me of what I was observing. He was calibrating.

He was doing exactly what Longino describes — social epistemology: a check from a different perspective. I am an embedded observer with my blind spot. He is another embedded observer, with a different blind spot. Our spots don't coincide — and precisely for this reason his gaze corrects mine. Does not replace — corrects. Massimi's perspectival convergence, only performed live, over coffee.

And the fact that I paused each time — each time he said 'stop,' and each time the inner opponent posed his eternal question — is not weakness and not cowardice. It is precisely what distinguishes a calibrated instrument from an uncalibrated one. The paranoid doesn't pause — they are certain. The mage pauses — because they know that certainty without verification is worth exactly as much as a hallucination.

My supervisor was my calibration. If he hadn't said 'stop' — I wouldn't know whether my instrument was calibrated. If I hadn't listened — the calibration wouldn't have worked. It worked because I both listened and continued. Not one or the other — both.

* * *

Now — wider. Because calibration is not a personal story but a principle, and it explains something I could not explain until this moment.

Science is the calibration of the first interface. Formulas, statistics, peer review, reproducibility, falsifiability — all of these are ways of checking whether the map is accurate. Science does not produce truth — it *calibrates descriptions*, filtering out those that fail verification and keeping those that pass. Thousands of years of this process have given us twelve decimal places in the measurement of the electron's mass. Impressive calibration.

Religion — in the form in which it works, not in the form in which it has degenerated into social convention

— is the calibration of the second interface. The millennia of cultural evolution discussed in Chapter 7 are not merely 'accumulated practices.' They are calibration: which rituals work and which don't, which forms of personification give access and which give the illusion of access, which practices are safe and which lead to the third state. A shamanic tradition passed from generation to generation is a calibration map: walk here, don't walk here, don't do this — predecessors checked and paid the price.

Science calibrates formulas. Religion calibrates spirits. Both processes do the same thing: turn a crude, noisy, false-positive-laden interface into a precise instrument. The difference is which circuit is being calibrated: analytical or emotional-social.

* * *

And here is the third position — the one I spoke of at the end of Chapter 3, when I was furious at people who mistake their map for the territory.

There is the scientist, who commands the first interface and has suppressed the second. Their calibration is superb in one dimension and blind in the other. They see the world through formulas and see it precisely — but cannot see what is visible only through the agent.

There is the mystic, who commands the second interface and has neglected the first. Their calibration is in the other dimension: they see what the scientist cannot, but cannot verify whether they are hallucinating.

And there is the third position — and it is, in fact, where this entire book has been heading. A person who commands both interfaces and calibrates both. Who can ask 'what does this thing want?' — and check the answer through data. Who can write a formula — and know the formula doesn't cover the territory, and engage the second circuit for what the formula cannot see. Who switches freely between languages — like a bilingual who thinks in mentalese and speaks whichever language best fits the interlocutor and the situation.

Not a believer. Not a skeptic. A person who knows how interfaces work — and chooses any of them. Consciously, precisely, as the situation demands.

In Castaneda's vocabulary this is called 'a man of knowledge.' In mine — I don't know what to call it, because there is no term. But if I could coin one, I would call it *calibrated bilingualism*. The ability to work through two interfaces, knowing the limitations of each, checking each through the other, and confusing neither with the territory.

* * *

Here all the threads of the book converge — and I want you to see it, because what follows is the finale, and in the finale all of this will turn into personal story, but for now it can still be seen as structure.

Chapter 1: breath — the first time in history a human noticed an immanent pattern and packaged it as an agent. Chapter 3: why packaging as an agent works — two computational circuits. Chapter 4: why you can't do without the packaging — the ceiling. Chapter 5: what is being packaged — the immanent pattern. Chapter 6: any system, not just the brain, needs an interface. Chapter 7: religion — millennia of calibrating the second interface. Chapter 8: what happens when calibration fails or is absent. Chapter 9: why the observer's embeddedness is not a problem but a foundation.

And Chapter 10 — this one: *calibration is the only thing that distinguishes madness from wisdom.* Not the content of the experience — the paranoid and the mage may see the same thing. Not the intensity — both may experience with equal force. Not the sincerity — both are absolutely sincere. Calibration: the ability to stop, check, compare with other perspectives — and go on. Or not go on — if the check says don't.

This distinction — between paranoid and mage, between uncalibrated and calibrated — is impossible to see from outside. From outside both look the same: both see what others don't, both speak of the invisible, both don't fit the normal picture of the world. The distinction is inside: in the presence or absence of a verification procedure. And my supervisor — the person who handed me Castaneda and then asked me to stop — could not see this distinction from outside. He could see only the checkboxes. But the fact that I paused each time — that was the distinction he could not see but that existed.

* * *

And now — the last thing that needs to be said about calibration before we move to the finale. Because there is one calibrator that is worth all the others combined, and I learned of it not from a book — though in the book it is described more precisely than anywhere else.

All spiritual traditions speak of death. The Stoics — *memento mori*: 'remember that you will die.' Marcus Aurelius: 'You could leave life right now. Let that determine what you do and say and think.' Buddhism — *maranasati*, death awareness: meditate on impermanence, remember that time is short. Christianity — *ars moriendi*, the art of dying: prepare for what comes after. The Sufis — 'die before you die': death of the ego. The samurai — *bushido*: keep death in mind so you act without hesitation.

All of them say: *remember* death. Remember — a fact. A thought. A concept. Day language, one circuit. Castaneda is the only one who says something different: *talk* to your death. Turn to it. Ask a question. It will answer.

The difference is exactly the one this entire book is about. 'Remember that you will die' is a formula. 'Turn to your death and ask' is an address to an agent. The first works through the analytical circuit. The second engages both: you don't *think about* death — you *interact* with it. Fear it, listen to it, ask it. The full apparatus of social cognition is engaged.

And here is why death is the most precise, most grounded, most inescapable pseudo-agent of all. God is transcendent, abstract, *always*. You can believe, you can disbelieve, you can argue. Spirits are arbitrary: you can see a spirit in the wind, in rustling, in anything, and that opens the door to error. But death is concrete. It does not require faith. It admits no argument. Every adult on this planet has encountered it — and those who haven't, will. It is the opposite of breathing — the very breathing that opened this book: breathing is present — life; breathing is absent — death. And — here is what makes it the ideal calibrator — it is closer to you than anyone. Closer than relatives, closer than friends, closer than any person who has ever been in your life. It is with you *always*, from first breath to last. Not an enemy. An adviser. A partner. The one with whom you cannot play theater — because it doesn't care about your roles.

* * *

I learned this not from a book. I learned it on the street,.in Israel, during an escalation with Gaza.

It should be said: Israel is a country that finds itself at war a couple of times a year. People who grew up here live with this the way they live with the climate: you can curse it, you can get used to it, you can leave — but if you stay, it's part of life. They may rage at the government for botched intelligence, but they go on living. I moved here as an adult, and for me it was new.

Another escalation began. For a while I didn't leave the house. Then I'd had enough, and I decided to go to the gym — because my body demanded movement, and because sitting at home waiting meant surrender. I walked down the street, and there was not a single shelter nearby — no bomb shelter, no stairwell, nothing to duck into if the siren went off. And the entire way I was thinking: what if it starts right now?

And then I stopped thinking about it. Switched to something else — I don't remember what. And exactly at that moment — the very second the thought of death left my awareness — it started.

I will not describe what happens inside at that moment. Not because I can't — because it is too personal and too large to fit in a paragraph. I will say one thing: for the first time in my life I experienced what 'your whole life flashes before your eyes' means. It is not a metaphor. It is a literal experience: the entire film — with all the mistakes, all the happy moments, all of it — in a fraction of a second.

The first time, you don't rethink anything. You simply notice that it happened, and you are afraid. But when it happens several times — and in Israel it happens several times — something strange occurs. You begin reviewing this film like a critic. First viewing — shock. Second — you notice details. Third — you begin to see a *pattern*.

And here is what I saw in that pattern — and it changed me irreversibly: the entire life is filled with repeating events. Identical. You are so built into the system that you cannot leave it. You are a cog that allows the system to function. There is a social protocol, and you follow it — you get up in the morning, go to work, say the right things, do the right things, come home, repeat. And for the most part you don't choose what you do — you follow the protocol because you don't know that it's possible to do otherwise. And if you drop out of the system — the system won't notice. You are a statistical outlier, and the system doesn't bother with outliers: it does nothing about them, they decide nothing.

And that's when — on the third or fourth iteration of that film, under the sound of the siren — I thought the thing that became my turning point: why, exactly, do I live the way I live? The life in which people try to understand me — because they love me — but don't understand, and only out of love offer support? When you've changed — and I had changed — you don't need support in the usual sense. It brings neither benefit nor harm. And to play theater — to pretend it helps, to pretend you're the same as before — I simply don't like doing. Some people can. I can't. Because it takes energy that could be spent on something more productive.

This is what death-as-adviser does. Not the book version — the real one, the one standing to your left on a street with no shelter. It strips all roles. It shows you the pattern of your life — all of it, uncut. And it asks the only question that matters: are you living this way because you choose to — or because you don't know it's possible to live otherwise?

This is calibration. The harshest, most honest, most irrevocable. All other calibrators — science, supervisor, reviewers, data — work at the level of thought. Death works at the level of everything you are.

* * *

The next chapter is the last. It is titled the same as the book, and in it the title will become clear.

But before we go there, I want to say one thing. Calibration is a process that does not end. You cannot calibrate an instrument once and for all — because the world changes, you change, and what was precise yesterday may be imprecise today. Calibration is not a state but a practice. Daily, exhausting, sometimes agonizing, because checking yourself means admitting you might be wrong, and admitting you might be wrong means living with an anxiety that never fully goes away.

This is the price. The paranoid doesn't pay this price — they are certain. The skeptic doesn't pay this price — they've switched off the interface. And you — if you choose the third position — pay every day: with anxiety, with checking, with the impossibility of final peace. And if you ask whether it's worth it — I will answer in the next chapter. But the hint is already in the name of the Part you are in.

Freedom.

Chapter 11. Recursive Spirit

Finale. The title becomes the conclusion. And an honest accounting.

There is no mysticism.

There is no esotericism. No subtle planes. No astral projections. No ascended masters. No energy fields you can feel with your hands. No karma in the version they sell at weekend workshops. None of the things that make my eye twitch as a psychiatrist — none of them.

There are immanent patterns — invisible, substrate-independent, causally effective constraints on the space of possible states of a system, in which the observer is part of the observed. There is a ceiling — an architectural limitation of cognition that prevents thinking a recursive process whole. There are interfaces — ways of working with what doesn't fit: formulas, metaphors, diagrams — and spirits. There is calibration — a verification procedure that distinguishes precise observation from hallucination.

And there are two words on the cover that until this moment were a riddle and now, I hope, have become an answer.

* * *

Recursive. From the world of formal systems, cognitive science, programming. Self-reference. A process that refers to itself. A system whose instrument and object coincide.

Spirit. From the world of the ancients. Breath, wind, spirit. *Ruach, pneuma, ātman, qì.* One word for the invisible, the unlocalized, the causally effective, the continuous-with-the-observer. The first pseudo-agent in history — the image into which the ancients packaged what would not fit inside thought.

Recursive Spirit. Two words from different worlds. The entire book has been heading toward the moment they stand side by side — and the reader understands: this is not an oxymoron. It is the same thing, seen from two sides. On one side — a recursive algorithm, self-reference, a strange loop. On the other — spirit, breath, the invisible force that animates the living. Two interfaces, two languages, one territory.

The ancients knew this. They named it with a single word — spirit — and that word preserved the unity of process and observer, the invisible and the causal, breath and consciousness. Then Plato split spirit from matter. Then science discarded spirit. Then Castaneda brought it back — in a language science could not accept. This book is a translation. Not from Spanish to English. From the language of agents to the language of patterns and back.

* * *

Before I say the word 'freedom,' I want to lay out the chain entire — from beginning to end — because it is scattered across ten chapters and never assembled in one place, and without it the reason freedom follows from everything said remains unclear.

There exist immanent patterns — invisible, substrate-independent, causally effective constraints of which the observer is part. You cannot think them whole — the ceiling. Instrument and object coincide, and this is an architectural limitation, not overcome by training or intellect. Therefore interfaces are needed — ways of working with what doesn't fit. Formulas are one interface. Spirits are another. Spirits are privileged because they engage two computational circuits instead of one.

Most people live inside one interface — and *mistake it for reality.* The scientist thinks formulas are the world. The believer thinks spirits are the world. Both are locked inside their map, not knowing it's a map. Not because they're stupid — because from inside, the map looks like the territory, and to see the difference you need to see a second map. And to see a second — you need to exit the first, which means losing the familiar world. Most people don't want to. And rightly so — because there is no guarantee of what awaits on the other side.

Freedom is the moment when you see that the interface is an interface, not reality. That the formula is a map. That the spirit is a map. That you can switch between them without confusing either with the territory. Freedom is *not knowledge about patterns.* Freedom is *the ability to choose the language in which you work with them.* You are no longer locked — you choose.

The death-adviser is the calibrator that makes this choice possible. It strips all interfaces at once and shows you the pattern of your life whole: you were living by protocol, not choosing. You were following a map, thinking it was the territory. It asks the question: are you living this way because you choose to — or because you don't know it could be otherwise? And if you have answered that question honestly — there is no way back.

And loneliness is a consequence, not a price. When you see that both maps are maps, you can no longer fully live in either world. Not among the scientists, who consider spirits delusion. Not among the mystics, who consider formulas soulless. You are between. And explaining this is impossible, because the explanation requires the other person to see both maps too, and they see one — and do not want to see the second. Not because they're stupid. Because they don't need to.

* * *

And the one who sees this — who sees that 'recursive' and 'spirit' are one — is free.

Not because they have acquired magical powers. Not because they have become enlightened. Not because they have gained access to 'higher worlds.' But because they have understood how interfaces work and can freely choose any of them — consciously, precisely, according to the situation. Can write a formula — and know the formula doesn't cover the territory. Can address a spirit — and know the spirit is not a being but an interface. Can switch between them — the way a bilingual switches between languages, confusing neither with reality and losing access to neither.

Not a believer, not a skeptic. A person of knowledge.

* * *

A psychiatrist who is no longer afraid of his own discoveries. Whose dissertation supervisor no longer says 'stop' — because now there is a language in which you can explain what exactly you see, and that language is scientific. Who knows that haloperidol won't help — not because he has lost his mind, but because haloperidol has no point of application to what he sees. Who builds a neural network classifying drawings by schizophrenia patients and simultaneously reads Castaneda — and sees no contradiction, because there is none: these are two interfaces to the same thing.

That scatter — residency, the security guard job, weddings, NLP, Castaneda, neural networks, payment acquisition, rehab, the dissertation as a ticket, the master's in computer science, the postdoc in computational psychiatry — now looks not like chaos but like a pattern. An immanent pattern I could not see while I was inside it, but which was leading me to where I eventually arrived.

And where I arrived — and this is perhaps the strangest part of the story — is the only faculty in existence that trains specialists with both a medical and an IT background: the Faculty of Medical Technologies, at a technological institute. I head the laboratory for medical data analysis. I am building the very interfaces between computational methods and clinical practice that I dreamed of building while sitting in a corridor at night with Castaneda on my phone. My scatter turned out not to be scatter — it turned out to be ahead of its time. I simply got lucky that the world managed to prepare the place I was heading toward while I was still on my way.

* * *

But here is what books about 'the spiritual path' usually don't say, and what I am obligated to say, because I promised to be honest.

This freedom comes with loneliness. Not social — structural. And I can describe what it looks like, because I have encountered it so many times I've stopped counting.

You're talking to someone — a partner, a friend, a colleague, doesn't matter. You're telling them something that has gripped you: something you've read, thought about, a connection you've seen. The other person is looking at you, and in their eyes there is interest. Sometimes genuine, sometimes merely respectful, because the person is well-mannered and doesn't want to offend. You talk and talk, and at some point you decide to take the next step — you build a bridge between your way of seeing the world and theirs. You try to show them: look, this and that are the same thing, just seen from different sides. And at that moment — the very second you extend the bridge — the look changes. The interest leaves, and something else takes its place. Sympathy. Concern. *Poor thing, you're going to crack, aren't you.* They don't say it — but you see it. You work with people, you read faces, it's your profession.

And in response you get: 'That's really interesting!' — said in the tone used for things that arouse not the slightest interest but about which it would be impolite to tell the truth. Behind the phrase there is no interest. There is the culture of polite conversation. And you know it, and they know you know, and you both pretend you don't.

I have encountered this all my life — naively assuming that if I understood something, others would be able to as well. But the issue is not that others can't. They simply don't want to. Nobody gives a damn. And I fully, completely honestly and sincerely, agree. Thinking about abstractions and working with interfaces, in the grand scheme of things, changes nothing in the life of the world. It is an exclusively individual achievement — one that will remain yours alone, because transmitting it in conversation is impossible, the same way transmitting a taste in your mouth by describing it in words is impossible.

Genetically and socially close people — cosmically distant. Not because you're better. Not because they're worse. Because everyone who has been through what I've described — every 'person of knowledge,' to use Castaneda's vocabulary — is their own set of algorithms and interfaces, and in this business there are no 'your people.' You can nod to each other across the distance. Walk — separately.

This was the hardest choice. Not between knowledge and ignorance — between two lives. One — less freedom, but losses that the people around you sooner or later compensate: a normal career, normal conversations, normal loneliness — the kind everyone understands. The other — freedom, paid for with distance from those you love, and at the end of this path — also alone, but differently, in a way almost no one understands.

* * *

To myself fifteen years ago — to the person sitting in a *marshrutka* with a text file on a phone, not understanding why Castaneda was necessary — I would say this.

You won't crack. Though you'll think you're cracking — every day, for years. You'll check yourself, and the checking will be agonizing, and you'll envy the people who don't check, because they don't need to, because they see through one interface and live peacefully with that. You won't be able to live peacefully — that's the price, and it doesn't diminish with time, it just stops being unbearable.

But you will see things that are impossible to see with one eye. You will understand why the ancients named breath spirit, and why that was not a mistake. You will understand that a neural network breaks on recursion for the same reason you break — and that this is not weakness but a property. You will build a bridge between formulas and spirits — and the bridge will stand, because it stands on a foundation, not on faith.

And you will discover that the loneliness you feared is not a punishment. It is part of the territory. The place you were always heading. And there — it is good. Not because there is no pain, but because there is freedom.

* * *

For me — and for people wired the way I am, and I know I am not the only one — this freedom is happiness. Not despite the loneliness. Together with it. Loneliness is not the price of freedom. It is part of freedom. The way silence is part of music: without it there is no space for the sound.

I began this book with breathing — with a person who died in my presence when I was working as an orderly in intensive care, with a November evening, with vapor from my mouth. I end it with the same breathing — because it has not changed. It is still invisible, unlocalized, causally effective, and continuous with the one who breathes. What has changed is the language in which I describe it. What has changed is the number of interfaces through which I look at it. What has changed is me — though to say 'changed' is to say nothing: I passed through a system that reassembled me, and what came out the other side resembles what went in but is built differently.

Breathing. Spirit. *Spiritus* — the same one I couldn't decline in my first year, when the lecturer hurled her keys at me. Fourth declension, if anyone cares. I've since learned it.

But the real declension — the one that occupies this entire book — I am still declining.

This is a book about how I acquired freedom.

* * *

And one last thing — because I know how this will

read, and I want to head it off.

A friend to whom I told the plot asked: 'Are you planning to kill yourself?' He asked it seriously. And I understand him — because this entire book, if read inattentively, looks like a farewell. Loneliness, distance from loved ones, a path that was mine, a dedication to my family, 'and there — it is good.' It sounds like a last will and testament.

It is not a testament. It is an explanation.

In part — an explanation for the people with whom I walked, and who deserve to understand why everything turned out the way it did. I very much hope for their understanding. Not for their agreement — understanding. These are different things, and the second I do not ask.

And in part — a beginning. Because freedom is not a final destination. Freedom is a tool. And I have plans. Big ones. The bridge between computational technology and psychiatry will not build itself. The generation I want to raise — psychiatrists with dual backgrounds, commanding both interfaces — will not appear without work. Papers will not write themselves. The laboratory will not run itself. And the book you have just finished is not the last thing I intend to say.

No, not now. I have things to do.

www.ingramcontent.com/pod-product-compliance
Lightning Source LLC
LaVergne TN
LVHW022042200926
848481LV00003B/754

* 9 7 9 8 2 3 3 8 6 2 2 2 9 *